100 Q&As on Carbon Peaking and Carbon Neutrality

Ying Chen and Qingchen Chao et al.

100 Q&As on Carbon Peaking and Carbon Neutrality

Translated by Pei Cai

PETER LANG

New York · Berlin · Bruxelles · Chennai · Lausanne · Oxford

Library of Congress Cataloging-in-Publication Data

Names: Chen, Ying, author. | Chao, Qingchen, author.
Title: 100 Q&As on carbon peaking and carbon neutrality / Ying Chen
and Qingchen Chao et al.
Other titles: Tan da feng, tan zhong he 100 wen. English
Description: New York : Peter Lang, [2024] |
Translation of: 碳达峰, 碳中和100问. |
Includes bibliographical references.
Identifiers: LCCN 2023035237 (print) | LCCN 2023035238 (ebook) |
ISBN 9781636674230 (hardback) | ISBN 9781636676777 (ebook) |
ISBN 9781636676784 (epub)
Subjects: LCSH: Carbon dioxide mitigation—China. |
Carbon dioxide—Economic aspects—China.
Classification: LCC TD885.5.C3 C436613 2024 (print) |
LCC TD885.5.C3 (ebook) | DDC 363.738/7460951—dc23/eng/20231018
LC record available at https://lccn.loc.gov/2023035237
LC ebook record available at https://lccn.loc.gov/2023035238
DOI 10.3726/b21254

Bibliographic information published by the Deutsche Nationalbibliothek.
The German National Library lists this publication in the German
National Bibliography; detailed bibliographic data is available
on the Internet at http://dnb.d-nb.de.

Cover design by Peter Lang Group AG

ISBN 9781636674230 (hardback)
ISBN 9781636676777 (ebook)
ISBN 9781636676784 (epub)
DOI 10.3726/b21254

This edition is an authorized translation from the Chinese language edition
Published by arrangement with Social Sciences Academic Press (China)
All rights reserved

WORLD ASSOCIATION FOR CHINA STUDIES (WACS) SERIES

© 2024 Peter Lang Group AG, Lausanne
Published by Peter Lang Publishing Inc., New York, USA
info@peterlang.com – www.peterlang.com

ADVISORS

Xiangwan Du, Academician of the Chinese Academy of Engineering, Honorary Director of the National Climate Change Expert Committee, Director of the Second National Climate Change Expert Committee, and Deputy Director of the Expert Advisory Board to the National Energy Commission

Yihui Ding, Academician of the Chinese Academy of Engineering, Deputy Director of the National Climate Change Expert Committee, and the first Director of the National Climate Center of the China Meteorological Administration

Dadi Zhou, Former Director general of the Energy Research Institute of the National Development and Reform Commission, Executive Vice Chair of the China Energy Research Society, Commissioner and Senior Research Fellow of the National Climate Change Expert Committee

Panmao Zhai, Co-chair of the Working Group I of the IPCC Sixth Assessment Report, Commissioner of the National Climate Change Expert Committee, and Research Fellow of the Chinese Academy of Meteorological Sciences

Yongsheng Zhang, Director general and Research Fellow of the Research Institute for Eco-civilization of the Chinese Academy of Social Sciences

LEAD AUTHORS

Ying Chen, Research Fellow of the Research Institute for Eco-civilization of the Chinese Academy of Social Sciences and Deputy Director of the Research Center for Sustainable Development of the Chinese Academy of Social Sciences

Qingchen Chao, Director general and Research Fellow of the National Climate Center of the China Meteorological Administration

Lei Huang, Research Fellow of the National Climate Center of the China Meteorological Administration

Mou Wang, Research Fellow of the Research Institute for Eco-civilization of the Chinese Academy of Social Sciences and Secretary General of the Research Center for Sustainable Development of the Chinese Academy of Social Sciences

Yongxiang Zhang, Associate Research Fellow of the National Climate Center of the China Meteorological Administration

Ying Zhang, Associate Research Fellow of the Research Institute for Eco-civilization of the Chinese Academy of Social Sciences

Kejun Jiang, Research Fellow of the Energy Research Institute of the National Development and Reform Commission

OTHER CONTRIBUTORS

Xianhong Lv, Lecturer of the School of Economics and Management of Tianjin Agricultural University

Wenmei Kang, Ph.D. candidate at the University of the Chinese Academy of Social Sciences

Hongren Su, MA student at the University of the Chinese Academy of Social Sciences

Linbo Pu, MA student at the University of the Chinese Academy of Social Sciences

Zhixuan Ji, MA student at the University of the Chinese Academy of Social Sciences

Ziyue Zhang, undergraduate student in Sociology, Tsinghua University

CONTENTS

FOREWORD CARBON PEAKING AND CARBON NEUTRALITY LEADS THE ENERGY REVOLUTION

On the occasion of the fifth anniversary of the Paris Agreement, China announced the national goal of achieving carbon peaking before 2030 and carbon neutrality before 2060. Based on scientific argumentation, this goal has become China's national policy and strategy to actively tackle climate change. Against such a backdrop, China must advance the energy revolution and make more steadfast and proactive efforts for low-carbon energy transition.

The energy transition is inevitable for continuous progress of human civilization. The discovery and utilization of fossil fuels, such as coal, oil, and gas, has dramatically raised labor productivity, thus fueling the shift from agricultural civilization to industrial civilization. This is the typical energy revolution, and it has greatly contributed to the progress of humanity. However, throughout the past 200 years and beyond, industrial civilization has posed severe risks to the environment, climate, and sustainability. The development of non-fossil fuel energy in modern times is advancing the shift from industrial civilization to ecological civilization as well as a new round of energy revolution.

It is noteworthy that China is different from other countries in terms of the three stages of the energy transition. In the first stage, coal was the dominant energy source globally. In 1913, it accounted for 70 percent of the world's primary energy. Several decades later, oil and gas became the dominant energy source globally. Now, the world is embracing the shift to the third stage. Likewise in the first stage, coal was the dominant energy source in China. But in the second stage, oil and gas did not become the dominant energy source in China, which instead has established a diversified, complementary, and balanced energy structure by using both fossil fuels and non-fossil fuel energy. In this way, China aims to phase in green, low-carbon, safe, and efficient energy transition, so as to embrace electrification, artificial intelligence, the Internet of Things (IoT), and low carbon emissions in the future. Now, we are entering the third stage, in which non-fossil fuel energy will be the dominant energy source.

Thanks to the rapid growth of the energy sector, China has enjoyed fast economic development since the reform and opening-up. Although remarkable achievements have been made in energy efficiency and energy restructuring, they can never be called revolutionary. Extensive economic growth featuring energy-intensive industries, low energy efficiency, and high-carbon energy structure has brought about increasingly acute environmental problems.

As China has listed energy intensity and carbon intensity as two major assessment indicators in recent years, the energy elasticity coefficient has seen a steady decline. However, China's energy intensity is still 1.3 times the world average, which is apparently unsustainable. If this figure can be lowered to 1.0, the GDP of the same scale can save more than one gigaton of standard coal.

Now we have only ten years left before achieving carbon peaking in 2030. Therefore, a detailed planning is extremely crucial for the energy sector during the 14th Five-Year Plan period because it will pave the way for China to achieve carbon peaking before 2030 and clarify pathways for China to achieve carbon neutrality before 2060.

During the 14th Five-Year Plan period, China needs to clarify requirements for energy conservation and energy efficiency, which should be the top priority of China's energy strategies, the first green and low-carbon energy, and the very element which guarantees the security of national energy supply and demand as well as the balance between energy and the environment. Especially given the current energy structure with fossil fuels as the dominant energy source, energy conservation and energy efficiency should be taken as the key solution to emissions reduction. In terms of energy production, we need the shift from black and high-carbon fossil fuels to green and low-carbon non-fossil fuel energy.

During the 14th Five-Year Plan period, the energy sector of China must embark on a new journey of high-quality development. When it comes to the development of fossil fuels, the energy transition should be taken into account.

For example, coal should be used in a clean and efficient manner, stabilizing oil production and increasing gas production should be the golden rule for the oil industry, and non-fossil fuel energy should be exploited to the fullest. To ensure that the elasticity coefficient of energy consumption is lower than 0.4 percent and energy consumption grows at an annual rate of 2 percent, China must rely on non-fossil fuel energy and natural gas as the main energy sources. Only in this way can China advance the energy revolution domestically, build a community with a shared future for humanity internationally, and foster a new development paradigm with domestic and international circulations reinforcing each other.

It is noteworthy that socioeconomic development in China has contributed to the rapid growth of renewable energy and drastic declines in the costs of energy production and energy storage. Between 2010 and 2019, the weighted average costs of photovoltaic power stations, solar thermal power plants, onshore wind projects, and offshore wind projects worldwide dropped by 82 percent, 47 percent, 39 percent, and 29 percent respectively. Ten years ago, renewable energy was nothing. But now, it has become something by shifting from a supplementary energy source to a mainstream energy source.

Take the distributed low-carbon energy network, for example. It can be used by households to generate electricity and to interact with centralized power grids. If such energy "producers and sellers" can be greatly encouraged in China, it will be easier to implement China's West-East Electricity Transfer Project and to transport coal from northern China to southern China. Three years ago, Lankao County in Kaifeng, Henan province, basically depended on coal from other provinces as its main source of electricity generation. But now, it is able to use its own energy network to generate electricity after three-year pilot energy revolution.

It is estimated that by 2025, with the development of hydropower, nuclear power, wind power, solar power, bioenergy, geothermal energy, energy storage technologies, new energy vehicles and other technical fields, and new business formats, such as integrated energy services, smart grids, microgrids, and virtual power plants, China's non-fossil fuel energy will account for 20 percent of primary energy; electricity will account for more than 30 percent of final energy; the share of the installed electricity generation capacity from non-fossil fuel energy sources will account for 50 percent, and the share of the electricity produced from non-fossil fuel energy sources will exceed 40 percent.

At that time, renewable energy will become the mainstream energy source and see a large increase during the 14th Five-Year Plan period. Without any increase in coal consumption, China is expected to reach peak coal and even achieve negative emissions from coal. It has been a clear-cut goal that regions/ cities in eastern China must take the lead in achieving carbon peaking before 2030 by increasing non-fossil fuel energy and re-electrifying vehicles (including electric vehicles) during the 15th Five-Year Plan period.

Carbon neutrality means having a balance between carbon emissions and carbon sinks after achieving carbon peaking. From this perspective, carbon dioxide is the main source of the world's greenhouse gas emissions, accounting for 73 percent. In 2019, global CO_2 emissions were 34.2 gigatonnes, followed by methane emissions. Since 2006, China has become the largest emitter in the world. After proposing the national strategic goal of achieving carbon neutrality, we have started to take more proactive steps towards the energy transition.

Considering the fact that fossil fuels are the dominant energy source, China and other countries can reduce CO_2 emissions from three aspects: Firstly, improving energy efficiency and reducing energy consumption. Especially when it comes to the fields of construction, transport, industry, and electricity, China should attach great importance to industrial restructuring and technological progress; Secondly, energy substitution. Non-fossil fuel energy, especially renewable energy, should account for a high proportion; Thirdly, carbon removal. China should increase carbon sinks and vigorously develop carbon capture, utilization and storage (CCUS).

Achieving carbon peaking and carbon neutrality remains a challenge for China. An unsuccessful transition will only lead to backward energy systems and technologies. But it is more of an opportunity from another angle. The energy transition will foster new industries, new growth points, and new investments, thus contributing to the sustainable development of economy, energy, the environment, and climate. Now, China is at a critical juncture in terms of the development of the energy sector and the times. We will encounter remarkable changes in future energy production, storage, and consumption, especially in the context of achieving carbon neutrality.

The great cause of achieving carbon peaking and carbon neutrality requires relentless efforts of the whole society. Ying Chen, Research Fellow of the Research Institute for Eco-civilization of the Chinese Academy of Social Sciences, Qingchen Chao, Director general of the National Climate Center, and the People's Daily Press, jointly planned the writing of the book *100 Q&As on Carbon Peaking and Carbon Neutrality* and compiled it. By explaining in simple language some core and key issues about achieving carbon peaking and carbon neutrality in the form of Q&As, this book will enhance the recognition and awareness of achieving carbon peaking and carbon neutrality among officials at all levels, enterprise personnel, and the general public. In my humble opinion, it is a very meaningful book, for which I strongly support its publication.

This article serves as the foreword for the book.

Xiangwan Du, Academician of the Chinese Academy of Engineering
April 16, 2022

PREFACE

In September 2020, China pledged to achieve carbon peaking before 2030 and carbon neutrality before 2060, which has re-kindled global enthusiasm to tackle climate change. Internationally, China has become an innovator and pacemaker in reducing CO_2 emissions. Domestically, all walks of life have taken proactive actions in response to the national strategy. Achieving carbon peaking and carbon neutrality has also become a buzzword for the media and been concerned by the whole society. The author panel have found in different situations and activities that people with different knowledge backgrounds have different understandings of the scientific connotations and policy significance of this concept, and of actions for implementing the concept. Coincidentally, the People's Daily Press also has intended to publish a book about these issues. All of these factors have contributed to the debut of this book.

Readers can know thoroughly from the book about topics on achieving carbon peaking and carbon neutrality. As for the format of this book, the first chapter clarifies the very significance of achieving carbon peaking and carbon neutrality to the great cause of building China into a great modern socialist country in an all-round way. Then the following chapters elaborate on "Where are we now?", "Where do we want to go?", and "How do we get there?". The second chapter introduces the background and scientific basis of achieving carbon peaking and carbon neutrality so that readers can know about impacts of climate change on natural ecosystems and human socioeconomic development as well as the necessity of achieving carbon peaking and carbon neutrality. The third chapter introduces policy actions to achieve carbon peaking and carbon neutrality, emphasizes the need for comprehensive socioeconomic transformation, and states suggestions on how to transform in various fields, sectors, and regions, as well as challenges and opportunities following transformation. The fourth chapter mentions the need for concerted efforts of the whole society, emphasizing that everyone can contribute to carbon peaking and carbon neutrality.

To know what readers are concerned about, the author panel designed multiple sets of questions and conducted a questionnaire via WeChat public account. Some 200 responses were collected. Among all the respondents, males accounted for 55 percent and females accounted for 45 percent, and 59 percent of respondents were under the age of 45 and 41 percent older than this age. It was also found that 69 percent of respondents worked for government institutions and public institutions, 17 percent worked in enterprises, 9 percent were students, and 5 percent had other occupations. Seventy percent of respondents were concerned about impacts and risks of climate change on China, the main factors affecting greenhouse gas emissions, challenges and opportunities for China and some key industries to achieve carbon peaking and carbon neutrality, and the role China plays in global climate governance. Sixty to Seventy percent of respondents were concerned about technologies, conceptual definitions, and international policy actions in specific industries. On average 55 percent of respondents were concerned about all the questions raised by the author panel. The questionnaire also included some open-ended questions so that readers could ask whatever they want to know. After analysis of the questionnaire, the author panel improved and adjusted relevant questions.

Launched by the Research Institute for Eco-civilization of the Chinese Academy of Social Sciences and the National Climate Center of the China Meteorological Administration, *100 Q&As on Carbon Peaking and Carbon Neutrality* was jointly compiled by some Chinese experts in relevant fields. The author panel referred to literature to elaborate on relevant issues in scientific and simple language. But it was so urgent and tough to complete the compilation of the book within the one-month time limit, from the date on which the task was assigned to the date on which the task was delivered (including China's Spring Festival). In addition, it was the first time for us to organize the compilation of this kind of book which involves multiple disciplines. Given these factors, we have to apologize that there may have some improper, inconsiderate, and even wrong treatments in this book for lack of experience. If any, do not hesitate to offer your opinions so that we can make revisions in the future.

In the process of compilation, experts from the Steering Committee provided thoughtful guidance, and editors from the People's Daily Press also gave many valuable suggestions. At the beginning of 2022, this book was funded by the Program of Translating (Chinese -English) and Publishing Academic Works on Innovation Engineering. Prior to the translation, minor revisions were made to the manuscript and the data were updated. We sincerely appreciate the great efforts made by all translators and editors from Social Sciences Academic Press.

The compilation and publication of this book was also jointly funded by the key research and development program of the Ministry of Science and Technology, "Research on Key Issues of Global Governance of and Domestic Response to Climate Change Risks" (2018YFC1509000), and the innovative engineering program of the Research Institute for Eco-civilization of the Chinese Academy of Social Sciences, "Research on the Strategy of Green Development in the Context of Achieving Carbon Peaking and Carbon Neutrality" (2021STSB01).

Ying Chen, Qingchen Chao
April 21, 2022

LIST OF FIGURES

LIST OF TABLES

· 1 ·

THE SIGNIFICANCE OF ACHIEVING CARBON PEAKING AND CARBON NEUTRALITY

It is necessary to identify the significance of achieving carbon peaking and carbon neutrality before taking concrete actions. Starting with the concepts of carbon peaking and carbon neutrality, this chapter analyzes the strategic significance of achieving carbon peaking and carbon neutrality for building China into a great modern socialist country in an all-round way and its close connection with the construction of ecological civilization, and focuses on major events set out in the 14th Five-Year Plan. After reading this chapter, readers can profoundly understand the scientific meaning of achieving carbon peaking and carbon neutrality as well as relevant policy actions.

1. What Is Carbon Peaking? What Is Carbon Neutrality?

Climate change is one of the most serious challenges for humanity. Since the industrial revolution, a large amount of carbon dioxide (CO_2) emitted through human activities, such as the burning of fossil fuels, industrial processes, and changes in agriculture, forestry and other forms of land use, has remained in the atmosphere and become the main cause of climate change. In addition to CO_2, other greenhouse gases responsible for the warming effect include methane, nitrous oxide, hydrofluorocarbons, perfluorocarbons, and sulfur hexafluoride. To tackle climate change and promote the sustainable development of human society, humanity must make relentless efforts to reduce greenhouse gas emissions.

When the volume of carbon emissions from global, national, urban, and corporate entities rises to the highest point, peak CO_2 emissions are achieved, after which emissions begin to fall. Most developed countries have achieved carbon peaking, and their CO_2 emissions have started to decrease. Although China's current emissions have increased at a slower speed, compared to the period between 2000 and 2010 which saw a fast increase, the growing trend has continued, and carbon peaking has not yet been achieved. Carbon neutrality refers to a state of balance between anthropogenic emission sources and anthropogenic carbon sinks through afforestation and carbon capture and storage (CCS). The goal of achieving carbon neutrality can be set at different levels, including the globe and different countries, cities, enterprises, and activities. In a narrow sense, achieving carbon neutrality only involves carbon dioxide emissions. But in a broad sense, it involves all greenhouse gas emissions. In terms of carbon dioxide, the concepts of carbon neutrality and net zero carbon emissions can be used alternatively. But when it comes to non-CO_2 greenhouse gases, the case becomes more complicated. Methane is a short-lived greenhouse gas. As long as its emissions are stable, net zero is not a must, and it will not affect the climate system in the long run.

According to the 2020 Global Carbon Budget released by the Global Carbon Project (GCP) in December 2020, it was estimated that continents and oceans around the world absorbed about 54 percent of global carbon dioxide emissions. So, will carbon neutrality be achieved if 50 percent of global carbon dioxide emissions are reduced? The answer is no. It is noteworthy that sinks for achieving carbon neutrality do not include natural carbon sinks or carbon stocks, but those increased for the sake of human activities, such as afforestation and forest management. For absorbing carbon dioxide, oceans have become increasingly acidified, thus adversely affecting marine ecosystems. Though terrestrial ecosystems are carbon neutral, they are not permanent carbon sinks. For example, though forests absorb carbon dioxide in most times, they absorb less when they become mature and emit carbon dioxide into the air after they die and decay. A forest fire may also quickly turn carbon stored in forests into carbon dioxide, which will be emitted into the air. In a word, carbon neutrality can only be achieved when anthropogenic emissions of carbon dioxide are removed by anthropogenic carbon sinks.

According to key findings of the Special Report on *Global Warming of 1.5°C* (SR15) approved by the Intergovernmental Panel on Climate Change (IPCC) in 2018, meeting the 2 degrees Celsius target under the Paris Agreement requires global CO_2 emissions to decline by about 25 percent from the 2010 level by 2030 and to achieve carbon neutrality around 2070; meeting the 1.5 degrees Celsius target requires global CO_2 emissions to decline by about 45 percent from the 2010 level by 2030 and to achieve carbon neutrality around 2050. Whatever the target, global CO_2 emissions are required to peak as soon as possible during 2020–2030.

Prior to the Paris Climate Conference in 2015, China pledged to peak emissions around 2030, and to cut carbon dioxide emissions per unit of GDP by 40–65 percent from the 2005 level and to increase the share of non-fossil fuels in primary energy consumption to about 15 percent, to increase forest cover by 40 million hectares above the 2005 level, and to increase forest stock by 1.3 billion cubic meters above the 2005 level by 2020. On September 22, 2020, in an address to the United Nations General Assembly, President Xi announced that China aimed to achieve carbon peaking before 2030 and carbon neutrality before 2060. On December 12, 2020, he further proposed new measures to scale up China's nationally determined contributions at the Climate Ambition Summit: by 2030, China will lower carbon dioxide emissions per unit of GDP by over 65 percent from the 2005 level, increase the share of non-fossil fuels in primary energy consumption to around 25 percent, increase forest stock by 6 billion cubic meters above the 2005 level, and bring the total installed capacity of wind and solar power to over 1,200 GW.

On December 18, 2020, the Central Economic Work Conference deployed the task named "Being Well-Prepared for Achieving Carbon Peaking and Carbon Neutrality" as one of the eight tasks in 2021, in a bid to achieve carbon peaking and carbon neutrality. To that end, a leading group in charge of the work related to carbon peaking and carbon neutrality was established at the central level, and it successively launched the "1+N" policy system. In this system, "1" refers to guiding opinions that sets out the overarching principles of all forthcoming policies; "N" includes action plans to achieve carbon peaking and carbon neutrality before 2030, implementation plans for keys fields such as energy, industry, urban and rural construction, transport, and agriculture and rural areas, as well as key industries such as the steel industry, the petrochemical industry, nonferrous metals industries, the building materials industry, the oil and gas industry, and supporting plans involving scientific and technological support, fiscal support, finance, the capacity of carbon sinks, data collection and calculation, supervision and assessment, etc. Various departments, such as the Ministry of Ecological Environment, the National Energy Administration, the Ministry of Industry and Information Technology, the National Development and Reform Commission, and the People's Bank of China, have all made a series of announcements to hasten actions to achieve carbon peaking and carbon neutrality. Achieving carbon peaking and carbon neutrality has become a new topic sparking heated discussions in the whole society.

2. What Factors Are Behind China's Proposal of Achieving Carbon Peaking and Carbon Neutrality?

Climate change is a major global challenge for humanity. Two factors are behind China's proposal of achieving carbon peaking and carbon neutrality in response to climate change. On one hand, China needs to achieve sustainable development, strengthen the construction of ecological civilization, and realize the goal of building a beautiful China. On the other hand, China needs to assume international responsibilities as a major country and contribute to building a community with a shared future. From September 22, 2020 in which President Xi delivered an address to the United Nations General Assembly to December 12, 2020 in which he delivered another important speech at the Climate Ambition Summit which marked the fifth anniversary of the Paris Agreement, President Xi announced twice within less than 100 days China's new goal of actively tackling climate change. His announcements have affirmed China's determination to follow the path of green and low-carbon development, drawn a blueprint for green, low-carbon, and high-quality development, and shown China's charisma of assuming international responsibilities as a major country, thus injecting robust political impetus into the implementation of the Paris Agreement, the process of global climate governance, and post-epidemic green recovery.

From a global perspective, 2020 can be seen as the first year of carbon neutrality. In this year, countries around the world have successively proposed the goal of achieving carbon neutrality while updating nationally determined contributions. This marks the start of the international process towards carbon neutrality and has left a profound and far-reaching impact on future global economy and international order. As a major country in the world, China must not hesitate to move forward. Instead, we must be proactive and strive to play a leading role.

3. What Are the Connections Between the Goal of Achieving Carbon Peaking and Carbon Neutrality and China's Two Centennial Goals?

The 19th National Congress of the Communist Party of China put forward the Two Centennial Goals, that is, to basically realize socialist modernization by 2035 and to build a great modern socialist country that is prosperous, strong, democratic, culturally advanced, harmonious, and beautiful by the middle of this century. It also divided the modernization process from the period between 2020 and the middle of this century into two stages. The construction of ecological civilization is crucial for the realization of the Two Centennial Goals and the sustainable development of the Chinese nation, and is inevitable for building a beautiful China. On May 18 and 19, 2018, President Xi stressed at the National Conference on Ecological and Environmental Protection that we should fundamentally enhance the quality of the environment and basically realize the goal of building a beautiful China by 2035 by hastening the construction of systems for ecological civilization; by the middle of this century, we should comprehensively improve material civilization, political civilization, spiritual civilization, social civilization, and ecological civilization, foster green development and lifestyles in an all-round way, achieve the harmony between humanity and nature, comprehensively modernize national governance systems and governance capacities in the field of environmental improvement, so as to realize the goal of building a beautiful China. Achieving carbon peaking and carbon neutrality is an integrated goal with two closely-connected stages. Achieving carbon peaking before 2030 as the first stage is consistent with the goals that must be achieved in the first stage of realizing socialist modernization by 2035 and building a beautiful China. It is a crucial sign that China basically realizes socialist modernization by 2035. Achieving carbon neutrality before 2060 as the second stage is consistent with the goal of holding the increase in global average temperature below 2 degrees Celsius above preindustrial levels and pursuing efforts to limit it to 1.5 degrees Celsius set out in the Paris Agreement, and the goals of building China into a great modern socialist country and building a beautiful China by the middle of the 21st century. It is an integral part of building China into a great modern country.

However, the two stages are different from each other in terms of nature. Achieving carbon peaking is a short-term specific goal, while achieving carbon neutrality is a medium and long-term goal. They are complementary.

Achieving carbon peaking and striving to lower peak carbon emissions as early as possible can leave more space and flexibility for achieving carbon neutrality. That is to say, the later we achieve carbon peaking and the higher the maximum emissions are, the greater challenges and pressure we will have to achieve carbon neutrality. Achieving carbon neutrality ahead of schedule needs more improvements in current policies after we achieve carbon peaking. We can never make it only by means of existing technologies and policy systems. What we need most is comprehensive and profound transformation of socioeconomic systems.

4. What Are the Connections Between Achieving Carbon Peaking and Carbon Neutrality and the Construction of Ecological Civilization?

Achieving carbon peaking and carbon neutrality and the construction of ecological civilization complement each other. For the unsustainability of the traditional pattern of industrialization, shifting from traditional industrial civilization to ecological civilization is inevitable, and it is the precondition for achieving carbon peaking and carbon neutrality. Reducing carbon emissions substantially and achieving carbon peaking and carbon neutrality are also crucial for promoting the construction of ecological civilization.

Based on the traditional pattern of industrialization, industrial civilization in the wake of the industrial revolution represents the great progress that has been made throughout human history. However, this development pattern, characterized by huge production and consumption of industrial wealth, is highly dependent on input of fossil fuels and material resources, which has inevitably resulted in huge carbon emissions, consumption of resources, damage to the environment, global climate change, and unsustainable development. All in all, reducing carbon emissions substantially and achieving carbon peaking and carbon neutrality are the two preconditions for solving global unsustainability and climate change.

On one hand, the construction of ecological civilization is the precondition for achieving carbon peaking and carbon neutrality. Carbon neutrality means that economic development and carbon emissions must be decoupled from each other to a large extent. This requires a fundamental change in the high-carbon development pattern, that is, a shift from high-carbon production and consumption of industrial wealth to moderate consumption of material wealth and low-carbon supply that meets overall demands of the people. To that end, the mass should profoundly change their values or views on the concept of "good life". The concept of ecological civilization, that is, lucid waters and lush mountains are invaluable assets, is a manifestation of the profound shift from traditional values and the traditional development pattern to the low-carbon development pattern.

On the other hand, reducing carbon emissions substantially and achieving carbon neutrality are crucial for the construction of ecological civilization. The shift from the traditional pattern of industrialization to ecological civilization and green development is a process of "creative destruction". In this process, new green supply and demand will emerge in markets and phase out non-green supply and demand. The announcement that China pledged to achieve carbon neutrality before 2060 and take vigorous actions to reduce emissions has thus set new constraints and market expectations for a hastened shift. Against such a backdrop, resources of the whole society will be effectively allocated towards green development, thus making green economy more competitive and accelerating the construction of ecological civilization.

5. How to Understand the Importance of Major Targets Set out in China's 14th Five-Year Plan to the Goal of Achieving Carbon Peaking and Carbon Neutrality?

As the Chinese proverb goes, a journey of a thousand miles begins with a single step. The 13th Five-Year Plan laid out the specific target that China should foster low-carbon development and effectively control the total volume of carbon emissions. It is the first time that control on carbon emissions has been mentioned in China's five-year plans. During the 13th Five-Year Plan period, China has made remarkable achievements in tackling climate change and green and low-carbon development. At the end of 2019, China's carbon intensity decreased by about 48.1 percent from the 2005 level, and non-fossil fuel energy accounted for 15.3 percent of primary energy. China has completed the targets of reducing carbon intensity by 40–45 percent and increasing non-fossil fuel energy to 15 percent by 2020 ahead of schedule, thus lying a solid foundation for fully implementing nationally determined contributions and achieving carbon peaking and carbon neutrality. On December 12, 2020, President Xi updated China's nationally determined contributions by 2030, which marks the refinement and implementation of the overall goal in concrete fields and a major step towards carbon neutrality. As general secretary Xi has often stressed, tackling climate change is not at others' request but on our own initiative. So far, China's green and low-carbon development has been on the fast track. The 14th Five-Year Plan period is a critical period overlapping three phases, that is, China's economy is now in a slowing growth phase, a painful structural adjustment phase, and a phase of absorbing the adverse effects of previous stimulus policies. It is also a key time window for achieving carbon peaking. During the 14th Five-Year Plan period, China needs to reduce energy consumption per unit of GDP and carbon dioxide emissions by 13.5 percent and 18 percent respectively, strike a balance between green and low-carbon development and high-quality development, coordinate domestic situations and international situations, organize the compilation of The Special Plan for Climate Change During the 14th Five-Year Plan Period, consider and formulate more detailed action plans for achieving carbon peaking, accelerate the construction of the national carbon market, actively participate in global climate governance, and mobilize the whole society to materialize the beautiful blueprint for achieving carbon peaking and carbon neutrality.

· 2 ·

THE SCIENTIFIC CONNOTATIONS OF ACHIEVING CARBON PEAKING AND CARBON NEUTRALITY

Achieving carbon peaking and carbon neutrality is one of the goals proposed to tackle climate change. To profoundly understand the goal, we need to first know about the scientific basis of climate change and clarify the scientific connotations of achieving carbon peaking and carbon neutrality. This chapter clarifies relevant concepts of climate change and summarizes key scientific conclusions about climate change, its impacts on natural ecosystems and sustainable development of human society and economy, challenges of human responses to climate change, and major ways and means for human beings to tackle climate change.

Section I: The Scientific Basis of Climate Change

6. How to Correctly Understand the Definitions of Climate Change?

Climate refers to the weather pattern in an area over a time. It is the mean or the statistical result of a weather pattern in a period of time and reflects the basic characteristics of an area, such as cold, warm, dry and wet. It is the result of interactions between the atmosphere, hydrosphere, lithosphere, and biosphere, and is shaped by atmospheric circulations, latitude, altitude, and surface morphology altogether.

Climate change refers to statistically significant changes in the mean and extreme values of climate. The rise and fall of the mean indicate the change in the mean state of climate. The increase in extreme values of climate indicates that the increase in climate instability and anomaly. The United Nations Intergovernmental Panel on Climate Change (IPCC) defines climate change as the variability in climate based on natural changes and human activities; the United Nations Framework Convention on Climate Change (UNFCCC) defines it as "a change of climate which is attributed directly or indirectly to human activity that alters the composition of the global atmosphere and which is in addition to natural climate variability observed over comparable time periods."

Climate change is a concept closely related to time frames. The connotations, manifestations, and main driving factors of climate change are different under different time frames. In terms of time frames and influencing factors of climate change, climate change can be generally divided into three categories, namely, climate change in the geological record, climate change in historical times, and climate change in modern times. The climate change more than 10,000 years before is the climate change in the geological record, such as glacial-interglacial cycles; the climate change since the emergence of human civilization (within 10,000 years) can be called the climate change in historical times; the climate change in the global instrumental record since 1850 is generally regarded as the climate change in modern times.

7. How Has Climate Changed in the Past Century?

The global climate has experienced a systemic change characterized by warming in the past century. In 2020, the average concentrations of CO_2, CH_4, and N_2O in the global atmosphere were recorded 413.2 ± 0.2ppm, 1,889 ± 2ppb, and 333.2 ± 0.1ppb respectively, increasing by 49 percent, 162 percent, and 23 percent respectively from preindustrial levels (1750) and reaching the highest levels in the past 800,000 years. Although CO_2 emissions from fossil fuels fell by about 5.6 percent in 2020 due to restrictions related to the COVID-19 pandemic, the rate of increase in CO_2 emissions between 2019 and 2020 was slightly lower than that that observed between 2018 and 2019 but higher than the annual average rate over the past decade. For the increase in major atmospheric greenhouse gas emissions, the effective radiative forcing reached 3.14 W/m² in 2019, which was significantly higher than that caused by natural factors, such as solar activities and volcanic eruptions, and was the most important factor in global warming.

The global climate system continued warming in 2021. The global average temperature was about 1.11 degrees Celsius above pre-industrial levels (the mean between 1850 and 1900), thus making 2021 the seventh warmest year in a row since 2015. In the past century, the global ocean surface temperature has increased by 0.89 degrees Celsius (ranging from 0.80 degrees Celsius to 0.96 degrees Celsius), and the global ocean heat content has continued to increase and accelerated significantly since the 1990s. Between 1993 and 2019, the rate of the global average sea level rise was 3.2 mm/year; between 1979 and 2019, the Arctic sea ice extent continued decreasing, with monthly September ice extent showing a decline of 12.9 percent per decade. Between 2006 and 2015, the rate of mass loss from global mountain glaciers reached 1,230 ± 24 gigatonnes/year, with the amount of mass loss increasing by about 30 percent compared with that between 1986 and 2005.

In the context of global warming, China's land surface temperature has continued increasing in the past 100 years, with a rising rate of 1.56 ± 0.20 degrees Celsius/100 years, significantly higher than that of the global average land surface temperature (1.0 degrees Celsius/100 years). Between 1951 and 2020, China's annual average surface temperature rose an average of 0.26 degrees Celsius every decade; the average temperature in northern China rose significantly faster than that in southern China; the average temperatures in winter and spring rose faster than those in summer and autumn. Between 1961 and 2020, the annual average precipitation in China fluctuated greatly from year to year; the annual precipitation in northeast China, northwest China, most areas of Tibet, and southeast China continued increasing significantly; the annual precipitation decreased progressively from the south of northeast China and parts of North China to most areas of southwest China.

8. What Are the Causes of Climate Change?

The causes of climate change can be divided into two categories: natural factors and human factors (Figure 1). Natural factors include changes in solar activities, volcanic activities, and variability within the climate system. Human factors include the increase in atmospheric concentrations of greenhouse gases caused by human burning of fossil fuels and deforestation, changes in the concentration of atmospheric aerosols, as well as land use and land cover changes, etc.

Since the industrial revolution, CO_2 emissions from the use of fossil fuels, such as coal and oil, have caused the increase in the concentration of CO_2 emissions in the atmosphere, and the greenhouse effect of greenhouse gases, such as carbon dioxide, has led to the warming of the climate system. So far, numerous scientific theories and simulation experiments have verified the correctness of the theory of the greenhouse effect. Therefore, only by considering effects of human activities can we simulate the trend in global warming in the past 100 years, and only by considering impacts of human activities on changes in the climate system can we explain changes in the atmosphere, oceans, cryosphere, and extreme weather and climate events. More observations and studies have further proved that greenhouse gas emissions caused by human activities are also the main reason for changes in global extreme temperature events and may be the main reason for the intensification of heavy terrestrial precipitation at the global scale. More evidence has also revealed that human activities have an impact on extreme events, such as extreme precipitation, droughts, tropical cyclones, etc. In addition, at the regional scale, human activities, such as land use and land cover changes or changes in the concentration of aerosols, can also affect changes in extreme temperature events, and urbanization may exacerbate warming in urban areas.

Human activities have also led to the increase in regional temperature and extreme events in China since the middle of the 20th century. In western China, human activities, including greenhouse gases, aerosol emissions, and land use change, are likely to be the main causes of the increase in land surface temperature. They may lead to the increase in the frequency, intensity, and duration of extreme high temperatures, the decrease in the frequency, intensity, and duration of extreme low temperatures, the increase in summer days and hot nights, and the decrease in frost days and freezing days in China. They may also increase the probability of high temperatures and heat waves and reduce the probability of low temperatures and cold waves in China.

The problems of China's observational precipitation data, such as limited observation time and space, poor quality of data, limitations of models and simulations, and huge impacts of internal variability on precipitation, have posed great uncertainties to studies of impacts of human activities on precipitation changes in different regions of China. Present studies have shown that human activities have an impact on the decrease in light rainfall and the increase in heavy rainfall in eastern China since 1950, but have little impact

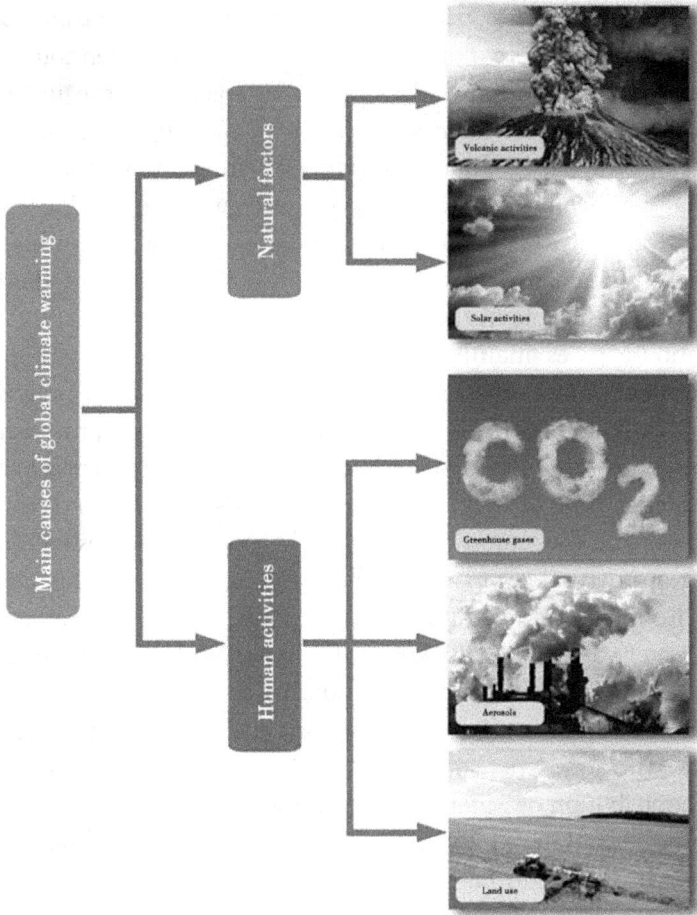

Figure 1. Main causes of modern global warming
Source: *Q&As on Climate Change*.

on the rainfall pattern (floods in the south of East Asian and droughts in the north of East Asian) caused by the East Asian summer monsoons. Since 1950, extreme precipitation in China has been increasing and intensifying significantly, indicating impacts of human activities to some extent.

9. Which Gases Are Defined as Greenhouse Gases?

In terms of the components of the Earth's atmosphere, nitrogen (N_2) accounts for 78 percent, oxygen (O_2) accounts for 21 percent, and argon (Ar) and other gases almost account for 0.9 percent. They account for more than 99 percent of gases in the atmosphere altogether. But they are not greenhouse gases. In general, these non-greenhouse gases hardly interact with the incident solar radiation and also basically do not interact with the long-wave infrared radiation emitted by the Earth. That is to say, they neither absorb nor emit thermal radiation, and basically have little impact on changes in the Earth's climate environment. It is the 0.1 percent greenhouse gases that have a huge impact on the Earth's climate environment. They can absorb and emit radiation, thus playing an important role in the Earth's energy budget.

Greenhouse gases mainly include water vapor (H_2O), carbon dioxide (CO_2), methane (CH_4), nitrous oxide (N_2O), ozone, carbon monoxide, and chlorofluorocarbons, fluorides, bromides, chlorides, aldehydes, nitrogen oxides, sulfides, and other trace gases. Water vapor can condense and precipitate. The typical residence time of water vapor in the atmosphere is ten days. The flux of water vapor into the atmosphere from anthropogenic sources is considerably less than from "natural" evaporation. Therefore, it does not contribute significantly to the long-term greenhouse effect. This is the main reason why tropospheric water vapor (typically below 10 km altitude) is not considered to be an anthropogenic gas contributing to radiative forcing. Anthropogenic emissions do have a significant impact on water vapor in the stratosphere, which is the part of the atmosphere above about 10 km. In terms of forcing and feedback, the contribution of water vapor in the stratosphere to global warming is much smaller than that of methane or carbon dioxide. Therefore, water vapor is generally considered to be a feedback medium rather than a forcing that causes climate change. Similar to the function of a greenhouse, carbon dioxide, methane, and other greenhouse gases can absorb long-wave radiation from the Earth's surface and have a positive effect on maintaining the suitability of the global climate. Without the greenhouse effect, the Earth's average surface temperature would be minus 19 degrees Celsius, rather than the current 14 degrees Celsius. However, a violent change in atmospheric concentrations of greenhouse gases in a short period of time will destroy the original stability and balance of the climate system.

Greenhouse gases can be basically divided into two categories. Some are inherent in the Earth's atmosphere and have significantly increased due to human activities since the industrial revolution (around 1750). They include carbon dioxide, methane, nitrous oxide, ozone, etc. Others are completely produced by human production activities (i.e., man-made greenhouse gases). They include chlorofluorocarbons, fluorides, bromides, chlorides, etc. For example, having been widely used in refrigerating machines and other industrial pro-duction, chlorofluorocarbons (such as CFC-11 and CFC-12) emitted though human activities have led to the destruction of stratospheric ozone. Since the 1980s, for the establishment of international conventions to protect the ozone layer, emissions of chlorofluorocarbons and other man-made greenhouse gases have been decreasing gradually.

10. How Does the Earth's Carbon Cycle Work?

The Earth's carbon cycle is a process in which green plants in the natural ecosystem absorb carbon dioxide from the air, convert it into carbohydrates and release oxygen through photosynthesis, and emit carbon dioxide into the atmosphere through biogeochemical cycles and human activities. Carbohy-drates can be stored in plants through photosynthesis and in animals through food chains. Plants and animals convert part of uptake into carbon dioxide and release it into the atmosphere through respiration, with the rest stored as the biological organism itself. After animals and plants die, most of them are finally emitted into the atmosphere in the form of carbon dioxide through microbial decomposition; only a tiny fraction of animal and plant residues are buried by sediments before that process and are converted into fossil fuels (coal, oil, natural gas, etc.) after a long time. When these fossil fuels are weathered or burned, the carbon inside is converted into carbon dioxide and released into the atmosphere (Figure 2)

Carbon is constantly moving between the atmosphere, oceans, and land in a cycle. Carbon dioxide can move from the atmosphere into oceans, and also from oceans into the atmosphere. These two forms of carbon exchanges both occurs at the junction of the atmosphere and oceans. Carbon dioxide in the atmosphere will become carbonic acid after being dissolved in rainwater and groundwater, then carbonic acid moves from rivers to oceans through run-off, and carbonate sediments form limestones, dolomites, and carbonaceous shales, etc. After chemical and physical processes, the carbon from these rocks is emitted into the atmosphere in the form of carbon dioxide. Human activi-ties release a large amount of carbon dioxide into the atmosphere when fossil fuels are burned. About 57 percent of carbon dioxide emissions are absorbed by natural ecosystems, and about 43 percent remain in the atmosphere, which causes the increase in the atmospheric concentration of carbon dioxide in the atmosphere from 280ppm before the industrial revolution to 410ppm in 2019 and the warming of the global climate system.

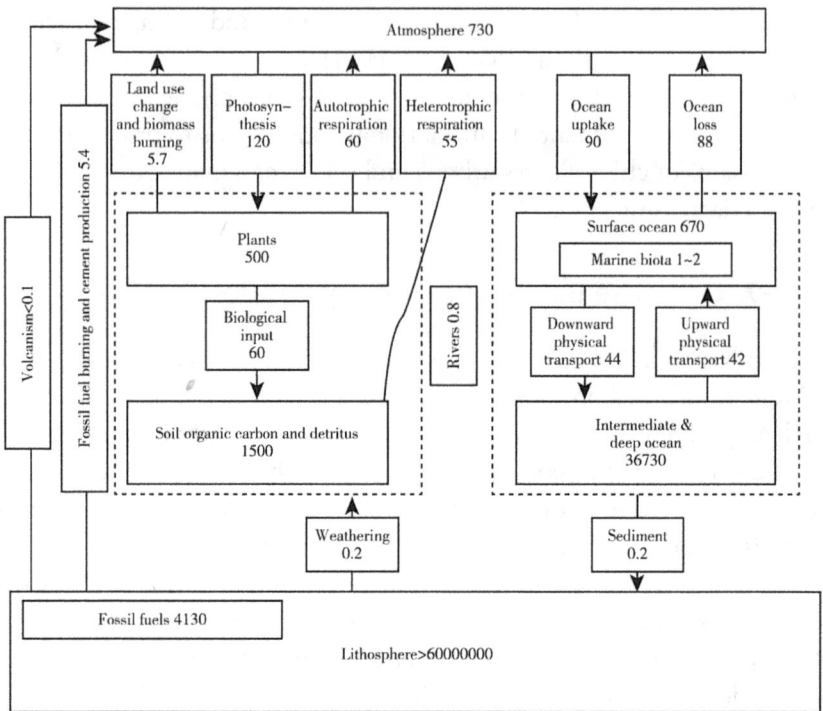

Figure 2. Schematic diagram of the global carbon cycle (Unit: gigatonnes/year, with arrows showing annual fluxes)
Source: IPCC, 2001; Falkowski, Scholes, Boyle, et al., 2000.

11. How to Measure Changes in Greenhouse Gases?

The increase in atmospheric concentrations of greenhouse gases has become the main cause of global climate and environmental changes. Measuring changes in the concentrations of major atmospheric greenhouse gases is crucial for us to better to study the sources, sinks, and transfers of these gases, to understand climate change, and to reduce energy consumption and air pollutant emissions. The three greenhouse gases of most concern are CO_2, CH_4, and N_2O. In the study of atmospheric chemistry, carbon monoxide (CO) also has a crucial impact on the greenhouse effect as an indirect greenhouse gas. Therefore, CO_2, CH_4, N_2O, and CO, are often taken as indicators to measure changes in the concentrations of major greenhouse gases. To measure greenhouse gases in the atmosphere, researchers normally first conduct field sampling and then send samples to laboratories for analyses. Carbon dioxide and methane are mainly analyzed through gas chromatograph with hydrogen flame ionization detector (FID), and nitrous oxide is mainly analyzed through gas chromatograph with electron capture detection (ECD). Other instruments, such as infrared spectrometers, can also be used to measure atmospheric concentrations of greenhouse gases.

To reflect background changes in atmospheric concentrations of greenhouse gases, researchers usually choose locations less severely affected by human activities to establish observation points for measurement. Data on global concentrations of greenhouse gases are usually obtained from the World Meteorological Organization Global Atmospheric Observatory (GAW), which includes 31 global atmospheric background stations, more than 400 regional atmospheric background stations, and more than 100 volunteer observing stations. In the early 1990s, China set up a global background station to observe concentrations of greenhouse gases at Waliguan in Qinghai province, and later established regional background observation stations at Shangdianzi in Beijing, Longfengshan in Heilongjiang province, and Lin'an in Zhejiang province. For example, the background observation station at Shangdianzi in Beijing can monitor changes in the concentrations of carbon dioxide, methane, and halogenated greenhouse gases directly, and can use other observation methods to monitor pollution sources and directions of greenhouse gases in the atmosphere, impacts of upstream cities on Beijing, impacts of Beijing on its downstream cities, and other relevant information.

There are only a limited number of ground observation points for monitoring the distribution of carbon dioxide concentrations, and these points are unevenly distributed. Such defects can be perfectly made up by satellite monitoring. Global satellite monitoring data as well as ground data and models can make the monitoring of concentrations of carbon dioxide and greenhouse gases more accurate.

12. What Is the Relationship Between Human-Induced Greenhouse Gases and Global Warming?

The relationship between greenhouse gas emissions and global warming is very complex because greenhouse gas emissions, concentrations of greenhouse gases, and global warming do not correspond with each other synchronously; the magnitude of global warming is approximately linearly correlated with cumulative global carbon dioxide emissions, which means that the greater cumulative global carbon dioxide emissions, the greater the magnitude of global warming. The IPCC Fifth Assessment Report (AR5) stated that holding the magnitude of global warming to less than 2 degrees Celsius relative to the period 1861–1880 with a probability of >66 percent would require cumulative global carbon dioxide emissions since 1870 to remain below about 1,000 gigatonnes of CO_2; holding the magnitude of global warming to less than 2 degrees Celsius relative to the period 1861–1880 with a probability of >33 percent would require cumulative global carbon dioxide emissions since 1870 to remain below about 1,600 gigatonnes of CO_2.

It should be pointed out that given the facts that the Earth's atmosphere itself contains a certain concentration of carbon dioxide, and that many natural ecosystem processes of the Earth absorb and release carbon dioxide, atmospheric concentrations of carbon dioxide vary over time and space. When carbon dioxide (be it natural or anthropogenic) is emitted into the atmosphere, it will be mixed by winds and distributed around the world over time. The mixing usually takes one to two months at the scale of the Northern Hemisphere or the Southern Hemisphere. It even takes more than a year at the global scale. Because the mixing between the Northern Hemisphere and the Southern Hemisphere is quite slow (atmospheric circulation of the Earth: the air moves across latitudes).

Here we use the volume of water in a swimming pool to represent the volume of carbon dioxide emission in the atmosphere, and use the changing water level to represent the changing volume of total carbon dioxide emissions in the atmosphere. Even in the absence of man-made carbon emissions, the water level of the pool will fluctuate naturally as rain (representing carbon dioxide emissions from the Earth's natural ecosystems) will raise the water level, and evaporation (representing carbon dioxide absorbed by the Earth's natural ecosystems) will lower the water level. If 10^{15} grams of carbon dioxide in the atmosphere are equivalent to 1 cubic meters of water, all carbon dioxide in the atmosphere can form a swimming pool with a length of 25 meters. a width of 15 meters, and a depth of 1.57 meters. In the absence of human-induced carbon emissions, each year 110 cubic meters of rainwater will flow into the swimming pool, and evaporation from the surface of the swimming pool will cause basically similar water losses. So, in the natural state, the water level of the swimming pool is basically stable (about 1.57 meters), in which case global warming will never occur. Since the industrial revolution, however, anthropogenic carbon emissions have increased fast as a faucet had been installed in the swimming pool. Here water flowing from the faucet into the swimming pool represents anthropogenic carbon dioxide emissions. The faucet adds about 10 cubic meters of water to the pool each year, but 5.7 cubic meters of water flows out for evaporation, with only 4.3 cubic meters of water left in the pool. That is to say, human-induced emissions have raised the water level by 11 millimeters (cumulatively 64 centimeters since the industrial revolution) to 2.21 meters. In other words, it is the 64-centimeter increase in the water level since 1750 that has caused the current global warming of more than 1 degrees Celsius above pre-industrial levels. In the future, the rising water level of the swimming pool will continue to cause global warming; only when the water level remains stable (net zero anthropogenic carbon emissions, or carbon neutral) will global warming stabilize at a certain level.

13. How Do Clouds and Aerosols Affect the Climate System?

Aerosols in the atmosphere are a multiphase (solid, liquid, and gas) system composed of atmospheric media as well as solid and liquid particles mixed in the atmosphere. They are the only non-gaseous component in the atmosphere and also a trace component in the atmosphere. Atmospheric aerosols mainly arise from human activities and natural emissions. Human-induced gases can be converted into aerosol particles through chemical or photochemical reactions. Aerosols in nature mainly comes from land surface, the atmosphere itself, and the injection from external space, of which land surface is the most important natural source. From the depths of the Earth's layers, some aerosol particles inject into the atmosphere through volcanic eruptions and can directly reach the stratosphere from about 15 km to about 50 km in altitude.

An important component of atmospheric aerosols is black carbon and organic carbon, which arise from fine particles and gaseous carbon compounds (later deposited on solid particles) emitted from incomplete combustion of fossil fuels. Black carbon aerosols strongly absorb solar radiation, ranging from visible light to near infrared light. Its absorption coefficient per unit mass is two orders of magnitude (100 times) that of dust. Therefore, although black carbon aerosol accounts for a small proportion in atmospheric aerosols, it has a great impact on regional and global climate.

Atmospheric aerosols can affect the Earth's climate by altering the radiation balance of the Earth's system. Studies have shown that aerosols have direct and indirect effects on climate. Direct effects occur when aerosols directly affect the radiation balance of the Earth-atmosphere system by scattering or absorbing short-wave and long-wave radiation. The magnitude of radiative forcing is closely correlated with optical properties as well as vertical and horizontal distributions of aerosols. Indirect effects of aerosols refer to the microphysical process in which aerosols change clouds in the atmosphere and lead to changes in radiation characteristics, cloud cover, and cloud lifetime, thus affecting the radiation balance of the Earth-atmosphere system and climate change. Clouds have an impact on the climate system in a very complex way. On one hand, clouds can reduce the amount of solar radiation absorbed by the Earth's surface by effectively reflecting solar radiation, thus cooling the Earth's surface. On the other hand, clouds can absorb short-wave radiation and produce long-wave radiation, thus warming the Earth's surface. Therefore, clouds and aerosols have a significantly uncertain impact on the climate system. Reducing uncertainties in these two aspects is very crucial for basic scientific research of climate change.

14. Will the Melting of Permafrost or Warming of Oceans Exacerbate Global Warming?

As an integral part of the cryosphere, permafrost is a permanently frozen layer of rocks and soils usually bound together by ice and remains at or below 0 degrees Celsius for at least two years. Frozen grounds account for about 50 percent of the land area on the Earth, and permafrost accounts for 25 percent of the land area. Studies have shown that by the end of the 21st century, the area of global permafrost will decrease by 40 percent even if robust actions are taken; the area of global permafrost will decrease by up to 80 percent if no actions are taken. The upper layer of permafrost is the active layer. Affected by global warming, permafrost warms and softens, and the active layer become thicker, which means that permafrost will become a huge carbon source as greenhouse gases (e.g. methane and carbon dioxide) in the thicker active layer are emitted into the atmosphere. In a word, greenhouse gas emissions from the active layer of permafrost will aggravate global warming. However, the changing mechanism of permafrost is very complex, and great controversies over the volume of greenhouse gas emissions from the active layer of permafrost, the emission rate, and regional differences still exist in the scientific community.

Oceans cover 71 percent of the Earth's surface, and 84 percent of them are over 2,000 meters deep. They play an important role in global climate change. As a crucial internal driving force of climate change, sea-land-atmosphere interactions play an important role in the transport and distribution of heat and water vapor at the global scale, and have an important impact on the global climate landscape and its evolution. For example, the tropical western Pacific Ocean has a warm pool that holds the warmest seawaters in the world. With the world's strongest convection and rainfall, it drives the Walker circulation and the Hadley circulation, thus regulating monsoons and El Niño. Due to the high heat capacity and huge mass, oceans have absorbed more than 90 percent of the excess heat associated with the increase in greenhouse gases since the 1950s, thus reducing the warming effect of greenhouse gases in the atmosphere to some extent. However, due to the long-term and large thermal inertia of oceans, they will always have a significant impact on global climate, even if anthropogenic greenhouse gas emissions can be reduced to net zero (carbon neutrality).

15. In Recent Years, Extremely Cold Weather Occurs Frequently in Many Areas of China in Winter. Is Global Warming Really Happening?

Between December 2020 and January 2021, cold air occurred frequently in China. The temperature in the northern part of northeast China and the northeast of Inner Mongolia even reached some -40 degrees Celsius degrees Celsius, and the observatory in the southern suburb of Beijing registered -19 degrees Celsius. Cold weather also continued to occur around the world. For example, in February 2021, many states in the United States were hit by extreme cold waves. In early March 2022, the snowfall in Istanbul hit the highest record in March since 1987, and the temperature reached the lowest in the past 30 years. As the saying goes, Rome is not built in a day. Likewise, the overwhelming trend in global warming cannot be changed by one or two cold waves. For example, in winter, the coldest area in China is the northern part of northeast China, in which the average temperature in January in the northern part of the Greater Khingan Range was as low as -30 degrees Celsius, and the Mohe Station even registered the lowest temperature of −52.3 degrees Celsius at on February 13, 1969, which was the lowest wintry temperature ever recorded in China. Much higher than the historical record, the Mohe Station registered −40 degrees Celsius in January 2021, and temperatures registered at other stations nearby were never lower than the historical record. Between January 4 and January 19, 1961, cold weather with the daily minimum temperature lower than −40 degrees Celsius lasted in Genhe of Hulunbuir city, for 16 consecutive days. Although such phenomenon occurred almost every winter throughout the 30 years between 1981 and 2010, it was incomparable to the historical extreme cold weather in terms of the degree and duration of coldness.

In January 2008, severe rain and snow disasters occurred in southern China, with the national average temperature (−6.6 degrees Celsius) 0.7 degrees Celsius lower than that in the same period (−5.9 degrees Celsius), hitting the lowest record since January 1986. However, it was still much higher than that in January 1977 and January 1955 (the national average temperature in January 1977 was close to −9 degrees Celsius, and in January 1955, it was lower than −8 degrees Celsius). In January 2008, the daily minimum temperature did not drop too low during the large-scale freezing rains and snows in southern China. For example, in January, minimum extreme temperatures lower than 10 degrees Celsius only occurred in 5 cities and counties of Anhui province, of which Dangshan had the lowest temperature of −12.2 degrees Celsius on January 29, followed by −11.7 degrees Celsius in Fuyang on January 31. However, before the 1980s, a cold wave would generally bring minimum temperatures along the Yangtze River down to −10 degrees Celsius. For example, a cold wave in January 1969 brought the minimum temperatures in Wuhan, Changsha, Nanjing, and Shanghai down to −17.4 degrees Celsius, −9.5 degrees Celsius, −13.0 degrees Celsius, and −7.2 degrees Celsius respectively. According to the historical records of minimum extreme temperatures at stations of these cities, the minimum extreme temperature in Wuhan could reach −18.1 degrees Celsius, that in Changsha could reach −11.3 degrees Celsius, and that in Hefei, located to the north bank of the Yangtze River, could be as low as −20.6 degrees Celsius. However, such low temperatures have never occurred since 1980s in the context of global warming.

From the point of natural phenomena, in the winter of 1977, the Dongting Lake, the Poyang Lake, the Taihu Lake, and other lakes along the Yangtze River were frozen for 7–10 days; in the winter of 1955, the Dongting Lake was also largely frozen, and the ice of the lake under the Yueyang Tower was one meter thick. However, since the 1980s, the Dongting Lake, the Poyang Lake, and the Taihu Lake have not been frozen ever, even in January 2008. Throughout the history, these large lakes and rivers in eastern China have been frozen in winter for many years. For example, in 1893, Shanghai was hit by a cold wave, and people could walk on the frozen Wusong River and the Taihu Lake; in 1862, the Huangpu River was frozen for half a month.

Why do some low temperature events still occur in the context of global warming? Climate change includes two aspects: changes in the mean state of climate elements in the global climate system; changes in the magnitude of climate change, namely changes in climate variability (increases in extreme weather and climate events). In the context of global warming, extreme weather and climate events have occurred frequently. Although warming events tend to be more frequent and intense on the whole, cold events, such as cold waves and extreme low temperatures, tend to decrease on the whole. This does not mean that low temperature weather will not occur in winter. Extreme cold events may still occur. Here we can make a metaphor. If we compare the global climate system to a pendulum, the swinging positions represent the degree of cold and warm events. In the context of global warming, the pendulum will swing more dramatically, and extreme hot events and extreme cold events may both occur. But extreme hot events may occur more frequently and intensely. That does not mean that we do not need to pay attention to extreme cold events which occur less frequently. We need to attach equal importance to them and proactively take precautionary measures.

16. How Do Scientists Predict Climate in the Coming Decades and Its Impacts?

To predict future trends and impacts of climate change, scientists typically use climate system models. The Earth system is composed of different spheres, including the atmosphere, lithosphere, hydrosphere, cryosphere, and biosphere. Interactions between the five spheres are a complex process. The process in which the Earth's climate system changes over time is affected by external forcings, such as volcanic eruptions and solar activities, as well as anthropogenic forcings, such as greenhouse gases emitted though human activities and land use change. Climate system models are a mathematical expression of the physical and dynamic process related to the momentum, mass, and energy of the Earth's climate system. Researchers can use it with a giant quantum computer to make quantitative and time-consuming calculations of big data involving complex evolution processes, to understand the evolution of the Earth's climate system, to simulate changes in external forcings and impacts of human activities, and to predict future trends in climate change. To project global and regional climate change, researchers also need to hypothesize future emissions of greenhouse gases and sulphate aerosols, namely emissions scenarios. Emissions scenarios are usually based on a number of factors (including population growth, economic development, technological progress, environmental conditions, globalization, the principle of equity, etc.). In recent decades, global climate system models have become increasingly complex and have evolved into

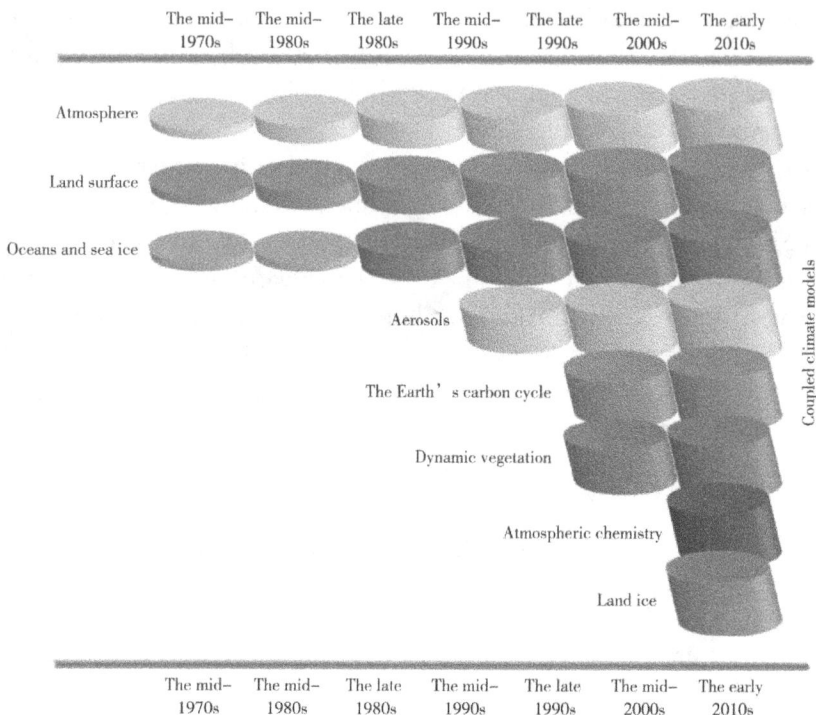

Figure 3. Schematic diagram of the development of climate models in recent decades
Source: Chapter I, the IPCC Fifth Assessment Report (WGI AR5).

Earth system models with the inclusion of biogeochemical cycles. Projections of future climate change are often based on the aggregation/integration of the same model or different models in different experiments. See Figure 3.

17. How Will Climate Change Evolve Without Any Human Control by the End of This Century?

The current global climate change is mainly attributed to greenhouse gases emitted though human activities. If human beings do not control greenhouse gas emissions, the Earth will continue to warm in the future, and this warming process will affect every aspect of the Earth. Scientists' projections of future climate have proven that the global average temperature by the end of this century would rise by about 4 degrees Celsius above pre-industrial levels, and the warming in the polar regions would be even higher. In the next 20 years, the global temperature will be more than 1.5 degrees Celsius above pre-industrial levels. The increase in the atmospheric concentration of carbon dioxide will acidify oceans. A warming of more than 4 degrees Celsius by 2100 will cause a 150 percent increase in oceanic acidity. Ocean acidification, global warming, overfishing, and habitat destruction have adversely affected sea creatures and ecosystems. A warming of more than 4 degrees Celsius by 2100 is likely to raise the sea level by 0.5–1 meter and even several meters in the next few centuries. Arctic sea ice may disappear completely through September. Climate change will have a serious impact on water supply, agricultural production, extreme temperatures, droughts, forest fires, and sea level rise. In the future, arid regions will become drier, and humid regions will become more humid. Extreme droughts may occur in the Amazon, western America, the Mediterranean, southern Africa, and southern Australia. Many places may incur higher economic losses in the future. Extreme events (such as large-scale floods, droughts, etc.) may cause malnutrition by affecting food production and increasing incidences of epidemic diseases. Floods can bring pollutants and diseases to a healthy water supply system, thus increasing incidences of diarrhea and respiratory diseases. Some species may extinct at a higher speed.

18. How to Measure the Space of Carbon Emissions?

No control on the current trend in global warming will only make the world warmer in the future and cause a warming of more than 4 degrees Celsius by the end of the 21st century. The Paris Agreement stated "holding the increase in the global average temperature to well below 2 degrees Celsius above pre-industrial levels and pursuing efforts to limit the temperature increase to 1.5 degrees Celsius above pre-industrial levels." Such statement was made on the basis of scientific and political decisions, believing that once global average temperatures exceed the threshold of 2 degrees Celsius warming, humanity may face greater dangers. To avoid such a potential risk, we must control greenhouse gas emissions within a certain range. In this sense, the space of carbon emissions refers to the range from the minimum volume of greenhouse gas emissions to the maximum volume of greenhouse gas emissions (within the limit on cumulative emissions) required to hold the increase in the global average temperature to a certain degree. It can be defined at global, national or sub-annual levels. Existing studies have suggested that the current anthropogenic carbon dioxide emissions are 42 gigatonnes per year, and that the remaining space of carbon emissions is less than 420 gigatonnes, which will be consumed within 10 years in the scenario of holding the increase in the global average temperature to 1.5 degrees Celsius. Therefore, nationally determined contributions proposed by countries under the Paris Agreement can hardly contribute to this scenario.

Section II: Impacts of Climate Change

19. Does Global Warming Do More Harm than Good or Vice Versa?

Climate change directly or indirectly affect humanity as well as human production and life (Figure 4). Current observations and scientific studies have suggested that global warming does more harm than good overall, but that it has different impacts on different regions and different industries. Take water resources, for example. precipitation changes and ice and snow melting in many regions have affected the quantity and quality of water resources. Glaciers continue to retreat in many regions, affecting downstream runoff and water resources; permafrost has also been warming and melting in high-latitude areas and high-altitude mountain areas. The world's major rivers are losing water. Geographical distribution, seasonal activities, migration patterns, and abundance of some biological species have changed over time. Climate change has done more harm than good to crop yield. Among all crops, wheat and maize have been more seriously affected than rice and soybean. Due to climate change, the yield of wheat and maize has decreased by about 1.9 percent and 1.2 percent per decade on average. Climate change may have adversely affected human health. Recent extreme weather and climate events, such as heat waves, droughts, floods, wildfires, and other climate disasters, have caused huge economic losses and casualties in many regions of the world. Global warming has caused sea level rise, damaging territories of some countries. Ocean acidification has led to more deaths of sea creatures.

20. Which Systems on the Earth Are More Sensitive to Climate Change?

Changes in some members of the Earth's climate system may mainly occur in certain areas, sometimes at an altitude of more than 1,000 km. They will affect the hemispheres and even global climate. Members that may change fundamentally are generally called members at a critical state and tend to be more sensitive to climate change. They have four characteristics: firstly, they have

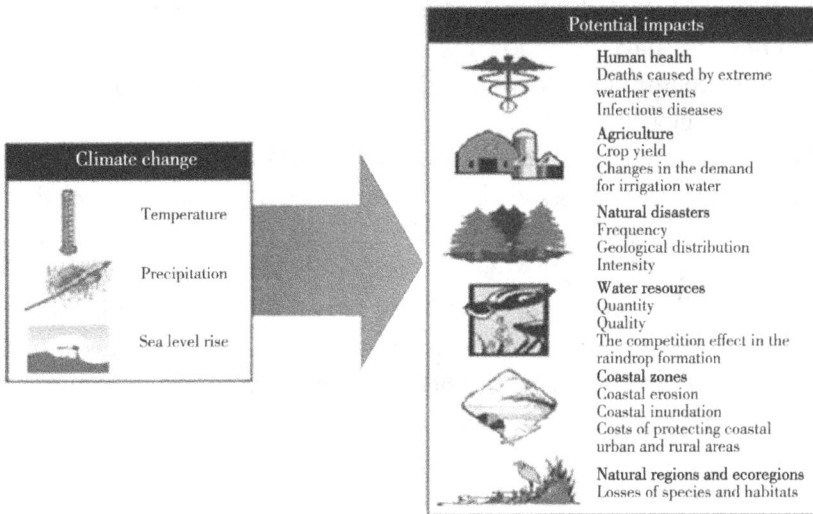

Figure 4. Impacts and risks of climate change
Source: The author made this figure based on available materials.

a threshold parameter; secondly, this parameter is related to climate change caused by human activities; thirdly, once the parameter reaches a tipping point, the state of these members will change fundamentally; fourthly, these changes will have a crucial impact on natural systems and socioeconomic systems. There are seventeen members sensitive to climate change in the Earth system. They are the Arctic summertime sea ice, the Greenland Ice Sheet, marine methane hydrates, permafrost, Himalayan glaciers, the West Antarctic Ice Sheet, the Atlantic Meridional Overturning Circulation, megadroughts in southwestern North America, the India Summer Monsoon, the West African Monsoon, the El Niño-Southern Oscillation (ENSO), forests of the Northern Hemisphere (North America), forests of the Northern Hemisphere (Eurasia), the Amazon rainforest, cold-water corals, tropical corals, the biological carbon pump in the Southern Ocean. Among them, the first six are climate elements of the cryosphere, the middle five are climate elements of atmospheric and oceanic circulation, and the last six are climate elements of the biosphere.

Variables, influence parameters, thresholds, and degrees of influence of these members have been partly understood. Take the Greenland Ice Sheet, for example. The main variable is ice volume, the influence parameter is temperature, the tipping point is 3 degrees Celsius, and the time frame of ablation is more than 300 years, which will raise the global sea level by 2–7 meters.

However, changing mechanisms of some members have remained unknow yet. For example, we have not known whether the ENSO will change more violently, or whether frequencies of El Niño or La Niña events will also change in the context of global warming.

So far, nine members sensitive to climate change have been activated, including frequent droughts in the Amazon rainforest, Arctic sea ice decline, slowdown in the Atlantic Meridional Overturning Circulation since 1950, fires and pests in the North American Boreal Forest, massive deaths of corals around the world, thawing of permafrost, faster melting of and ice loss from the Greenland Ice Sheet and the West Antarctic Ice Sheet, and faster melting of the East Antarctica Ice Sheet. These sensitive members are correlated with each other, and their activation will trigger positive feedback mechanisms of the greenhouse effect. Ice melting reduces the Earth's reflectivity, leading to rises in surface temperatures, sea level rise, deaths of sea creatures, and disruption of oceanic and atmospheric circulation patterns, which will in turn affect global temperatures and rainfall. Climate change may cause forests to die and thus produce huge greenhouse gas emissions, triggering the conversion of many systems on the Earth from carbon sinks to carbon sources. Once these sensitive members broke through the tipping points, a series of cascading effects will be triggered, thus exacerbating climate change, pushing more sensitive systems to cross the tipping point, and accelerating threats to human survival and civilization.

21. Has Global Warming Seriously Affected the World and China?

Climate change has left about 3.3–3.6 billion people in highly vulnerable environment globally. If average global temperatures exceed 2 degrees Celsius warming, 3–18 percent of species in terrestrial ecosystems may face a high risk of extinction, and direct losses caused by floods may be several times higher than ever before. Even if average global temperatures exceed 1.5 degrees Celsius warming, some ecosystems in the polar regions, alpine regions, and coastal regions will face irreversible effects.

China is part of the regions that are sensitive to global climate change and have been significantly affected. Since the 1950s, temperatures in China have been increasing faster than the global average. Climate change has exerted widespread impacts on China's natural ecosystems and human society. Frequencies of extreme weather and climate events in China have been on the rise. Extreme high temperature events, such as floods, urban waterlogging, typhoons, and droughts, have been more frequent, thus incurring more economic losses. Direct economic losses caused by extreme weather and climate disasters in China have increased by 1.4 times from an average of 120.8 billion yuan per year before 2000 to an average of 290.8 billion yuan per year after 2000. Climate change has also caused serious water problems in China. Runoff of major rivers in eastern China has seen declines. The runoff of the Haihe River and the Yellow River has decreased by more than 50 percent, resulting in the intensified contradiction between the decreasing supply and increasing demand of water resources in northern China. For lack of water resources, an increasing number of cultivated lands have been affected by droughts. Climate change has affected structures, functions, and services of China's ecosystems to varying degrees. Together with natural disturbances and human activities, it has led to less biodiversity and ecosystem stability as well as more vulnerability, thus making agricultural production more instable, costly, and lower-quality. In addition, sea level rise has aggravated coastal erosion, seawater (salt tide) intrusion, and soil salinization, and high sea levels superimposed by typhoon-storm surges have seriously affected the development of coastal cities. Extreme weather and climate events have exerted a significantly adverse impact on the operation of infrastructure and major projects. Increasingly frequent and serious climate risks have been threatening the stability of human systems and will have a cascading effect on China's sustainable development through complex economic and social systems. All in all, climate change has exerted widespread impacts on China's natural ecosystems and human society.

22. What Is the Relationship Between Climate Change and Changes in the Environment?

Climate change and the environment are correlated with each other in two aspects: firstly, climate change has led to ecological destruction and environmental degradation; secondly, changes in the environment have led to changes in the Earth's climate system. On one hand, fast warming of the Earth's climate system over the past century has exerted widespread impacts on natural ecosystems, including impacts on water resources and quality, glacier retreat, changes in geographical distribution, seasonal activities, migration patterns, and abundance of biological species, impacts on food production and quality, and impacts on human health. On the other hand, carbon on the Earth is mainly stored in the atmosphere, oceans, and land. Without the influence of human activities, the carbon budget between the atmosphere, oceans, and land would have been basically balanced on the whole. However, since the industrial revolution, human production and life have caused great damage to lands and ecosystems in the form of huge energy consumption, substantial use of fossil fuels, land reclamation, and urban development, and emitted many types of greenhouse gases (e.g. carbon dioxide and methane) into the atmosphere, thus causing fast warming of the Earth's climate system. Carbon emissions are the same cause of both climate change and environmental degradation. Therefore, proactive response to climate change will be conducive to environmental improvement; environmental protection and the construction of a beautiful ecology will also be conducive to the stability of the Earth's climate system. They are closely correlated with each other.

23. What Impacts Has Climate Change Exerted on Cities?

Climate change has exerted serious impacts on urban ecosystems, atmospheric environment, human health, and urban infrastructure, especially on the development of coastal urban agglomeration, in the form of the urban heat island effect, the urban dry island effect, the urban moisture island effect, the rain island effect, and the fog island effect.

In general, temperatures in urban areas change more violently than those in surrounding suburbs. In this sense, a city resembles a heat island. The urban heat island effect is usually accompanied by the urban dry island effect. As the main body of cities, urban underlying surface is made of continuously reinforced concrete so that it is impervious and tends to form a dry island separate from surrounding areas. Some cities are highly humid during a certain period of time. For example, in Shanghai, the moisture island effect occurs in January more frequently but less intensely, and it occurs in summer less frequently but more intensely. Poor air circulation caused by high-rise buildings in big cities as well as strong heat flow caused by air conditioners and vehicle exhaust emissions may cause more heavy rainfall events and even regional waterlogging in cities. The fog island effect is mainly caused by excessive urban particulate matter and cloud condensation nuclei. In recent 50 years, urban areas of China have seen an overall increase in smogs.

Climate change has exerted widespread negative impacts on urban infrastructure, residents, ecosystems, and economic systems. It has affected infrastructure systems, such as water and energy supply systems, sewerage and drainage systems, transportation and telecommunications, services including health care services and emergency response services, the built environment, and ecological services in urban areas. High temperatures and heat waves, rainstorms, snowstorms, and typhoons have damaged transportation equipment and ground facilities, posed great risks to traffic safety, and acutely affected the normal operation of urban roads, railways, aviation, and navigation. Climate change has affected the temperature of urban underlying surface, the height of offshore sea levels, and urban precipitation. So, in urban construction, site selection, drainage facilities, road planning, warehousing, and emergency response need to be adjusted in accordance with climate change, and planning and design standards need to be changed accordingly. Coastal urban agglomerations are vulnerable to climate change. Sea level rise and seawater intrusion will cause coastal erosion and soil salinization, and make rivers salty. Seawater intrusion will cause siltation of channels and abandonment of ports. Coastal cities will see a significant increase in hazards and risks related to flooding. In recent years, extreme events, such as typhoons, storm surges, and rainstorms, have occurred frequently in coastal cities in southeast China, causing numerous casualties and huge economic losses. Children, the elderly, and very vulnerable populations are the most vulnerable groups in urban areas.

24. Will Northwest China Turn into a Green Oasis from a Dry Land as It Has Become Warmer and Wetter in the Past 40 Years?

The annual precipitation in most parts of northwest China is less than 200 mm, and the precipitation in many parts of Gansu, Qinghai, and Xinjiang is even less than 50 mm. In the past 60 years, the annual average temperature in northwest China has increased at a rate of about 0.30 degrees Celsius/10 years, 2.5 times higher than the global average (0.12 degrees Celsius/10 years) and significantly higher than the national average (0.24 degrees Celsius/10 years). The annual precipitation in most parts of northwest China has been on the rise overall, but it differs greatly in different places. The annual precipitation in the central and western parts of northwest China has shown a significant upward trend, while that in the eastern part of northwest China (eastern Gansu, Ningxia, and Shaanxi) has shown a downward trend. In the central and western parts of northwest China, the climate has become warmer and wetter since 1987, and the precipitation has increased significantly since 2000, especially in spring. Climate change has led to faster glacier melting in northwest China and larger areas of lakes and wetlands, which is conducive to vegetation growth. But it will cause more hazards and risks.

The precipitation in northwest China is far less than the potential evaporation. In the context of climate change, precipitation may increase indeed, but very insignificantly. Because the potential evaporation in northwest China will also increase. Even if the amount of increased precipitation exceeds the amount of increased evaporation, it does not mean that climate conditions have changed fundamentally. It just indicates the alleviation of drought conditions. According to climate monitoring, semi-arid areas and wetter areas in northwest China have not seen a dramatic decrease, and extremely arid areas have seen a slight decrease and turned into semi-arid areas since 1961 (Figure 5).

Jiangnan (a geographic region to the south of the Yangtze River) has a monsoon-influenced humid subtropical climate. It has four seasons, abundant sunlight, agreeable temperature, abundant rainfall, humid air, and heavy rains and high temperatures in the same season. The annual precipitation in this region is more than 1,000 mm and can be up to 2,000 mm sometimes. Therefore, the arid climate in northwest China is not likely to be changed, and northwest China is not even possible to turn into a green oasis from a dry land, though it has become warmer and wetter in the past 40 years.

25. Does Climate Change Trigger Chain Reactions in Countries Around the World?

Climate change may trigger chain reactions in a region, a country, and often many countries on a large scale, especially in the context of socioeconomic globalization. For example, the global food crisis between 2007 and 2008 was triggered by successive droughts in Australia, a major bread basket in global wheat markets. In 2006, a drought known as the Millennium Drought occurred in Australia, and successive droughts in the wake of it caused continuous wheat yield losses. Their impacts also spread to livestock production and biofuel production as the inventory of the global food system had already been insufficient. In response to the global food shortage, governments around the world acted quickly. Six of the world's top 17 wheat exporters and four of the top nine rice exporters imposed varying degrees of trade restrictions. Global food supplies were thus sharply reduced, pushing food prices to soar accordingly. In high-income countries, food expenditure accounts for a relatively small proportion of total expenditure, but the case is contrary in low-income countries.

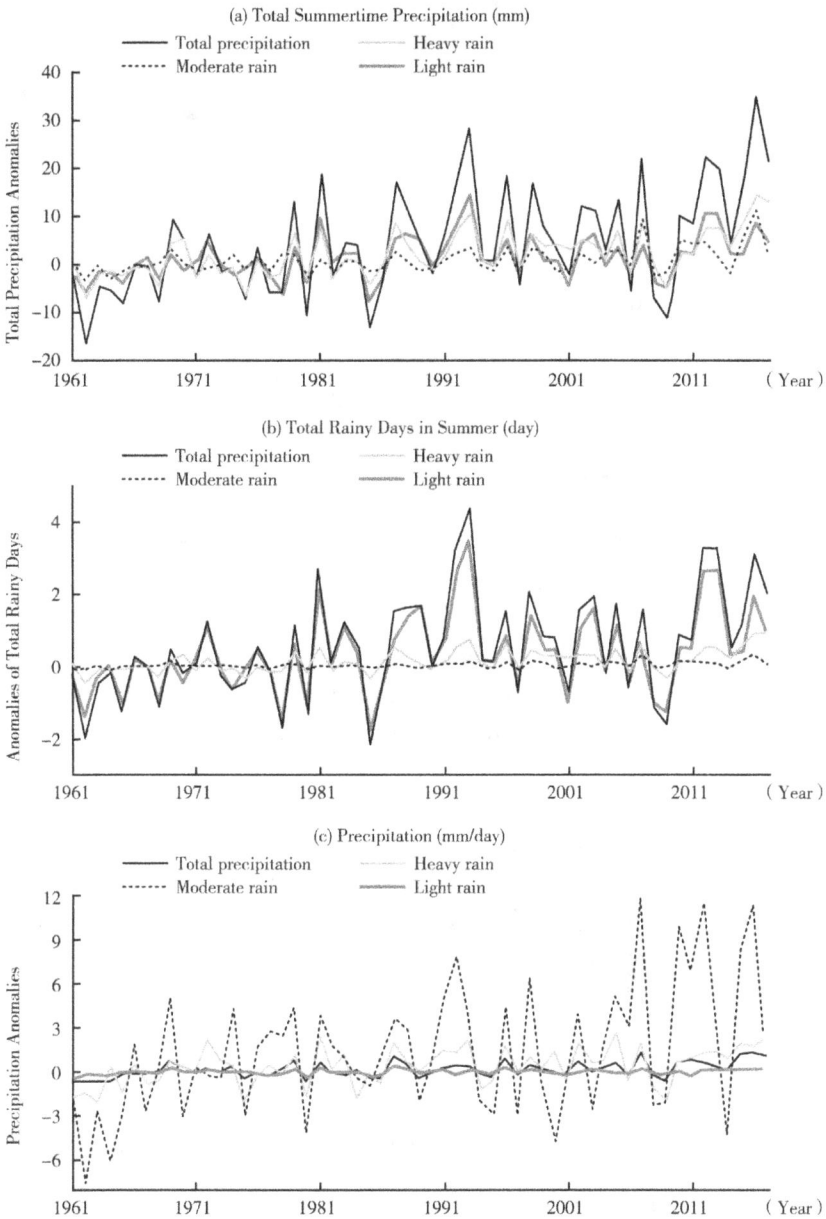

Figure 5. Total summertime precipitation between 1961 and 2017 in northwest China (35°–50° N, 72°–105° E)
Source: National Climate Center of the China Meteorological Administration.

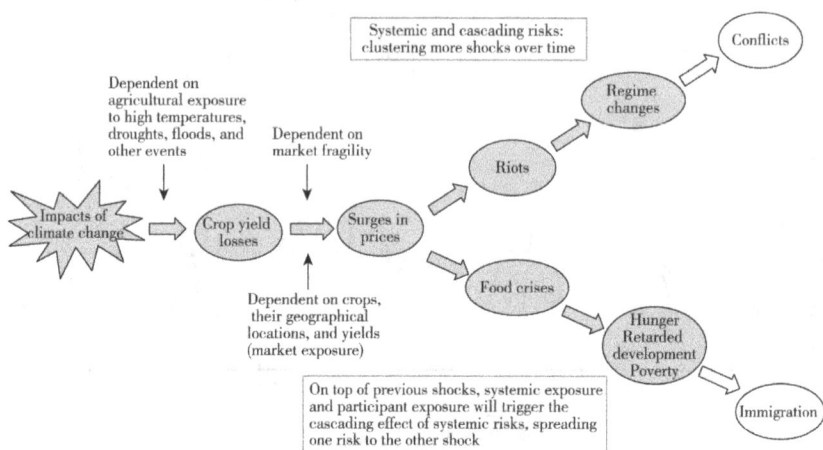

Figure 6. Conceptual framework for systemic risks induced by climate crisis
Source: *UK-China Cooperation on Climate Change Risk Assessment: Developing Indicators of Climate Risk.*

Countries that are heavily dependent on food imports are particularly vulnerable to surges in prices. For shortage of food, many countries have been affected by riots, and some countries have even suffered from regime changes throughout the history.

The probability that such climate risks affect a wide range of countries and industries may not be high. But once they occur, they will evolve and trigger chain reactions in an unpredictable manner. They are often triggered by a certain extreme event and will then affect structures, functions, and stability of more systems through a series of causal chains of risks, thus bringing about widespread and far-reaching consequences (Figure 6).

26. How Will Climate Risks Evolve in the Future?

Global climate change may trigger risks in terms of the following aspects.

Water resources: Risks in this aspect will increase significantly as concentrations of greenhouse gases increase, and renewable surface water and groundwater resources will decrease significantly in many arid and sub-tropical regions in 21st century, thus worsening inter-sectoral competition for water resources. The population affected by losses of global water resources will increase by 7 percent every 1 degree Celsius warming.

Ecosystems: Frigid Arctic tundra and the Amazon rainforest may face high risks; some terrestrial and freshwater species may be at higher risk of extinction.

Food production and food security: Some regions may face a warming of 2 degrees Celsius or more than 2 degrees Celsius compared to the late 20th century if they cannot adapt to climate change. It is estimated that climate change will only benefit a small number of regions; it may adversely affect yields of major crops (wheat, rice, and maize) in tropical and temperate regions.

Coastal systems and low-lying regions: They will be more vulnerable to inundation, coastal floods, and coastal erosion caused by sea level rise, and coastal ecosystems will suffer more pressure.

Human health: climate change will affect human health by exacerbating existing health problems, thus worsening people's health conditions in many regions, especially in low-income developing countries.

Economic sectors: a warming of around 2 degrees Celsius may trigger an annual economic loss of 0.2–2.0 percent to the income in most economic sectors.

Urban and rural areas: Many global risks occur in urban areas, and rural areas are more subject to risks in the aspects of water scarcity, food security, and agricultural income. Overall, the world will suffer moderate to high risks in the case of a warming of 1.5 degrees Celsius or 2 degrees Celsius above pre-industrial levels and will suffer high or very high risks in the case of a warming of 4 degrees Celsius.

27. Will Impacts of Climate Change Be Any Different in the Case of a Warming of 1.5 Degrees Celsius or 2 Degrees Celsius?

In 2021, the global average surface temperature was more than 1.11 degrees Celsius above pre-industrial levels. But different parts of the Earth had different degrees of warming. Some regions might have a warming of more than 2 degrees Celsius or less than 1.2 degrees Celsius. In the case of a warming of 1.5 degrees Celsius, the temperature increase in mid-latitude regions will be about 3 degrees Celsius in extreme hot days; in the case of a warming of 2 degrees Celsius, this figure will be about 4 degrees Celsius. In the case of a warming of 1.5 degrees Celsius, the temperature increase in high-latitude regions will be about 4.5 degrees Celsius at extreme cold nights; in the case of a warming of 2 degrees Celsius, this figure will be about 6 degrees Celsius. It is estimated that in the case of a warming of 2 degrees Celsius, some high-latitude regions and/or high-altitude regions of the Northern Hemisphere, eastern Asia, and eastern North America may face higher risks caused by heavy precipitation events as well as more frequent heavy precipitation related to tropical cyclones. Consequently, a larger proportion of the global land area will be affected by flooding caused by heavy precipitation.

A warming of 1.5 degrees Celsius will bring about numerous risks to and impacts on terrestrial and marine ecosystems, human health, food and water security, socioeconomic development, etc. However, its adverse impacts on natural and human systems are less significant compared to the case of a warming of 2 degrees Celsius. For example, at least one sea-ice-free Arctic summer is expected every 10 years for global warming of 2 degrees Celsius, with the frequency decreasing to one sea-ice-free Arctic summer every 100 years under 1.5 degrees Celsius; by 2100, global mean sea level rise is projected to be around 0.1 meter lower with global warming of 1.5 degrees Celsius compared to 2 degrees Celsius, preventing nearly 10 million people from being threatened; the degrees of ocean acidification and threats to coral reefs are also projected to be lower compared to the case of a warming of 2 degrees Celsius. Climate-related risks to health, livelihood, food security, water supply, human security, and economic growth are projected to increase in the case of a warming of 1.5 degrees Celsius and to be even higher in the case of a warming of 2 degrees Celsius (Table 1).

Section III: Main Ways to Tackle Climate Change

28. How Can Humanity Tackle Climate Change?

Humanity can tackle climate change through mitigation and adaptation. Mitigation refers to the control of greenhouse gas emissions and/or enhancement of greenhouse gas sinks through a variety of economic, technological, and biological policies, measures, and instruments. To prevent climate change from threatening ecological systems, food production, and sustainable socioeconomic development, and to limit atmospheric concentrations of greenhouse gases to a level in which the Earth's climate system is free from dangerous anthropogenic interferences, humanity must control or reduce greenhouse gas

Table 1. Risks in the cases of a warming of 1.5 degrees Celsius and a warming of 2 degrees Celsius

Fields	Risks in the case of a warming of 1.5 degrees Celsius	Risks in the case of a warming of 2 degrees Celsius
High temperatures and heat waves (the probability that the global population experience once every five years)	14%	37%
One sea-ice-free Arctic summer (the probability of one sea-ice-free Arctic summer)	at least once every 100 years	at least once every 10 years
Sea level rise (the sea level rise in 2100)	0.40 meters	0.46 meters
Extinctions of vertebrates (the probability that vertebrate species extinct by half)	8%	16%
Extinctions of insects (the probability that insects extinct by half)	6%	18%
Ecosystems (the proportion of the land area where biological communities change in the global land area)	7%	13%
Permafrost (the area of melting Arctic permafrost)	4.8 million km^2	6.6 million km^2
Crop yields (the yield of maize in the tropics declining by)	3%	7%
Coral reefs (declining by)	70–90%	99%
Fisheries (marine fisheries production declining by)	1.5 million tonnes	3 million tonnes

Source: The IPCC Special Report on *Global Warming of 1.5 degrees Celsius* (SR15), 2018.

emissions through policies and measures aiming to mitigate climate change. To control greenhouse gas emissions, we need to change energy structure, contain the use of fossil fuels, and increase the share of nuclear energy and renewable energy; we need to improve the efficiency of the power generation sector and other energy conversion sectors; we need to improve energy efficiency of the industrial sector and reduce energy consumption of a product; we need to improve household energy efficiency, such as home heating; we need to improve energy efficiency in the transport sector; we should try our best to cause less damage to forest vegetation and control methane emissions of paddy fields and landfills. Only by taking these measures, can we effectively control and reduce emissions of carbon dioxide and other greenhouse gases. To have greenhouse gases better absorbed, we need to rely on afforestation and adoption of carbon sequestration technologies, which refer to the process of separating and capturing carbon dioxide from burning gases, and then disposing it in deep seas and underground, or sequestering carbon dioxide in chemical, physical, and biological ways. Seen from the policy measures adopted by governments around the world, feasible ways that can directly control carbon dioxide emissions include limiting the use of fossil fuels, greenhouse gas emissions, and deforestation. Economic instruments can also contribute to tackling climate change. Governments can levy pollution taxes and fees, implement emissions trading (including joint compliance of countries), grant subsidies and development assistance; They can also encourage public participation by providing information to the public and developing leading-edge power generation technologies and other forward-thinking energy technologies for the 21st century.

Adaptation is the adjustment in natural or human systems in response to actual or expected climatic stimuli or their effects, which moderates harm or exploits beneficial opportunities. There are many ways to adapt to climate change. Adaptation measures include institutional measures, technological measures, and engineering measures for climate change adaptation, such as building infrastructure to tackle climate change, establishing monitoring and early warning systems for extreme weather and climate events, and strengthening management of climate disaster risks. In terms of agricultural climate change adaptation, farmers can better cope with droughts by adopting new drought-tolerant crops, intercropping, crop residue retention, and weed management, and developing irrigation technologies and hydroponic farming; they can better cope with flooding by making polders, improving drainage, developing and encouraging alternative crops, adjusting planting and harvesting times; they can better cope with heat waves by developing new heat-tolerant crops, changing planting schedules, and monitoring crop pests.

29. How Can Each Industry Reduce Emissions?

Climate change mitigation measures refer to economic, technological, and biological policies and instruments to control greenhouse gas emissions and/or enhance greenhouse gas sinks. Among all measures, reducing greenhouse gas emissions is at the core, and major transformation of the energy supply sector is a crucial guarantee for greenhouse gas emissions reduction.

> **The energy supply sector:** Power generation devices should be decarbonized. The share of zero-carbon or low-carbon energy supply from renewable energy, nuclear energy, and fossil fuels that will be treated with carbon capture and storage (CCS) should be greatly enhanced in primary energy supply, so as to crowd out coal-fired plants that do not adopt carbon capture and storage.

> **The field of energy use:** The energy sector should reduce energy demand through energy-efficient technologies, better modes of transport, behavioral changes, infrastructure improvement, and urban development; the building sector should reduce energy use by adopting new technologies, knowledge, and policies, and formulating energy efficiency policies as well as building codes and standards; the industrial sector should improve the existing energy efficiency, reduce emissions per unit of energy, recycle materials, and reduce product demand by upgrading products and adopting optimal technologies.

> **The field of agriculture and forestry:** Afforestation, reducing deforestation, and sustainable forest management are three effective ways to reduce emissions. Cropland and pasture management and soil restoration are the most effective ways to reduce emissions in agriculture. Urbanization boosts income growth but causes high energy consumption and high emissions. Governments should improve energy efficiency and land planning and contain expansion via cross-departmental collaboration to achieve emissions reduction.

> **Cross-sectoral collaboration:** The energy supply sector and end-use sectors are highly interdependent with each other in terms of emissions reduction. They should implement systemic and cross-sectoral strategies

together as early as possible to reduce costs and to improve effectiveness of emissions reduction.

As an important market instrument for emissions reduction, carbon emissions trading refers to initiatives that put an explicit or implicit price on carbon emissions to encourage producers and consumers to invest heavily in products, technologies, and processes with low greenhouse gas emissions. It is conducive to carbon emissions reduction.

30. What Are Negative Emissions Technologies?

To achieve carbon neutrality, we need to adopt negative emissions technologies to remove carbon dioxide emissions from the atmosphere and store them, so as to offset the carbon emissions that are difficult to reduce. There are two main types of carbon dioxide removal (CDR): 1. The nature-based approach: using biological processes to increase removal of carbon dioxide emissions and store them in forests, soils, or wetlands. 2. The technology-based approach: removing carbon dioxide emissions directly from the air or controlling natural processes of carbon removal to accelerate carbon storage. Examples of some negative emissions technologies are listed in Table 2. The mechanisms, characteristics, and maturity of different technologies are quite different.

In the short term, the nature-based approach of carbon removal can greatly help improve the quality of soils and water and preserve biodiversity. However, in the long run, it can hardly remove carbon dioxide emissions from the atmosphere once and for all. A forest fire may release carbon stored back into the atmosphere. We will also encounter numerous challenges to apply negative emissions technologies (e.g. BECCS and DACCS) on a large scale. For example, BECCS requires massive production of bioenergy, thus putting pressure on land and water resources; it involves production, collection, storage, transport, and utilization of bioenergy, as well as carbon capture, transport, and storage, etc. A detailed life-cycle assessment of this technology needs to be made to judge whether it can achieve negative emissions or not.

Table 2. Examples of negative emissions technologies

Technologies	Descriptions	Mechanisms of carbon dioxide removal	Methods of carbon sequestration
Afforestation/ reforestation	Sequestrating atmospheric carbon in living organisms and soils through afforestation	Biological	In soils/ vegetation
Biochar	Converting bioenergy to biochar and using biochar as a soil conditioner	Biological	In soils
Bioenergy with carbon capture and storage (BECCS)	Plants absorb carbon dioxide from the air and take it as bioenergy, and the carbon dioxide emissions from this bioenergy conversion is captured and stored.	Biological	In deep geologic formations
Direct air capture and carbon storage (DACCS)	Using engineering methods to capture carbon dioxide emissions directly from the atmosphere and store them	Physical/chemical	In deep geologic formations
Enhanced weathering/ Mineral carbonization	Enhanced mineral weathering allows atmospheric carbon dioxide to react with silicate minerals to form carbonate rocks	Geochemical	In rocks
Improving planting methods	Adopting no-till farming to enhance soil carbon sequestration	Biological	In soils
Ocean fertilization	Adding iron to oceans to enhance biological carbon sequestration	Biological	In oceans
Ocean alkalinity	Increasing ocean alkalinity to enhance carbon sequestration through chemical reactions	Geochemical	In oceans

Source: the author made this table based on available materials.

31. What Are the Pathways for Limiting Global Warming to 2 Degrees Celsius or 1.5 Degrees Celsius?

Carbon dioxide can exist in the atmosphere for up to 200 years. Even if humanity stops emitting carbon dioxide into the atmosphere, residual carbon dioxide in the atmosphere will still cause continuous global warming. The warming effect of residual carbon dioxide emissions in the atmosphere is known as the cumulative effect of carbon dioxide. Therefore, in considering the pathways for limiting global warming to 2 degrees Celsius or 1.5 degrees Celsius, scientists must take into account the remaining space for global emissions as well as the cumulative effect of carbon dioxide. The IPCC made a comprehensive assessment on the pathways for limiting global warming to 2 degrees Celsius or 1.5 degrees Celsius and listed GHG emission pathways under different levels of global warming after comparing and calculating the results of different models. The IPCC Fifth Assessment Report (AR5) stated that limiting global warming to 2 degrees Celsius requires global greenhouse gas emissions to decline by 40–70 percent from the 2010 level and emissions levels to reach near zero or below by the end of this century. In 2018, the IPCC Special Report on *Global Warming of 1.5 degrees Celsius* (SR15) stated that limiting global warming to 1.5 degrees Celsius requires global greenhouse gas emissions to decline by 40–60 percent from the 2010 level and emissions levels to reach near zero by 2050. The IPCC Sixth Assessment Report (AR6), released in 2022, stated that modelled global pathways that limit warming to 1.5 degrees Celsius with no or limited overshoot requires global greenhouse gas emissions to peak between 2020 and 2025 at the latest.

32. On What Scale Should Negative Emissions Technologies Be Applied to Achieve Global Carbon Neutrality?

The maximum temperature reached is determined by cumulative net global anthropogenic CO_2 emissions up to the time of net zero CO_2 emissions and the level of non-CO_2 radiative forcing in the decades prior to the time that maximum temperatures are reached. Take limiting global temperature increase to 1.5 degrees Celsius above pre-industrial levels, for example. Many possible emission pathways are available for achieving this goal. Emission pathways that limit warming to 1.5 degrees Celsius with no or limited overshoot requires global net anthropogenic CO_2 emissions to decline about 45 percent around 2030 from the 2010 level. To that end, a package of mitigation measures should be taken, including reductions in energy and resources intensity, decarbonization rates, and reliance on negative emissions technologies, such as carbon dioxide removal. By comparing the following four pathways, we intend to indicate different characteristics of different measures, and different scales and degrees of applying negative emissions technologies under different pathways (Figure 7; Table 3).[1]

> **Pathway 1 (P1):** A scenario in which social, business and technological innovations result in lower energy demand up to 2050 while living standards rise, especially in the global South. A downsized energy system enables rapid decarbonization of energy supply. Afforestation is the only CDR option considered; neither fossil fuels with CCS nor BECCS are used.
>
> **Pathway 2 (P2):** A scenario with a broad focus on sustainability including energy intensity, human development, economic convergence and international cooperation, as well as shifts towards sustainable and healthy consumption patterns, low-carbon technology innovation, and well-managed land systems with limited societal acceptability for BECCS.
>
> **Pathway 3 (P3):** A middle-of-the-road scenario in which societal as well as technological development follows historical patterns. Emissions reductions are mainly achieved by changing the way in which energy and products are produced, and to a lesser degree by reductions in demand.
>
> **Pathway 4 (P4):** A resource- and energy-intensive scenario in which economic growth and globalization lead to widespread adoption of greenhouse-gas-intensive lifestyles, including high demand for transportation fuels and livestock products. Emissions reductions are mainly achieved through technological means, making strong use of CDR through the deployment of BECCS.

---

1 IPCC, 2018: Global Warming of 1.5°C. An IPCC Special Report on the impacts of global warming of 1.5°C above pre-industrial levels and related global greenhouse gas emission pathways, in the context of strengthening the global response to the threat of climate change, sustainable development, and efforts to eradicate poverty [Masson-Delmotte, V., P. Zhai, H.-O. Pörtner, D. Roberts, J. Skea, P.R. Shukla, A. Pirani, W. Moufouma-Okia, C. Péan, R. Pidcock, S. Connors, J.B.R. Matthews, Y. Chen, X. Zhou, M.I. Gomis, E. Lonnoy, T. Maycock, M. Tignor, and T. Waterfield (eds.)]. In Press, p. 14.

33. How Can Climate Change Affect International Investments?

Since the late 1980s, climate change has exerted a profound impact on human survival and development. Greenhouse gas emissions causes global warming, environmental crises, and severe challenges to human beings. To effectively tackle climate change and promote international cooperation, the international community has signed a series of international agreements, such as the United Nations Framework Convention on Climate Change, the Kyoto Protocol, and the Paris Climate Agreement. To fulfill obligations of international climate agreements, governments have taken a lot of measures to encourage investments and development in low-carbon fields and formulated many measures to restrict or even cancel investments in fields with high-energy consumption, high emissions, and high pollution. These measures have incubated a new-type low-carbon economy, leaving foreign investors who are unable to meet the characteristics of a low-carbon economy, the standards for greenhouse gas emissions, and requirements for emissions reduction in a less competitive advantage.

To tackle climate change, international markets classify investment into climate-friendly investment and non-climate-friendly investment. Climate-friendly investment will contribute to greenhouse gas emissions reduction and implementation of climate change measures. On the contrary, non-climate-friendly investment may hinder actions on climate change. These two types of investment involve different policies and treatment. Host countries often take a series of incentives, such as tax incentives, to encourage climate-friendly investment, which will thus enjoy preferential treatment from host countries in the same field. Take automakers, for example. New energy vehicle manufacturers with low carbon emissions will enjoy more preferential treatment than traditional diesel engine automakers with high carbon emissions. Climate change may change a country's comparative advantages, thus changing capital flows and affecting international investment landscape, especially when the country has a comparative advantage in climate and geography. For example, if all fossil fuel power plants are required to adopt carbon capture and storage, they will have to face higher operating costs in investment; if carbon-intensive industries are required to implement strict emission standards, they will have to bear losses in profits and competitiveness.

As the main body of international investments, multinational corporations may help host countries realize low-carbon transition through foreign

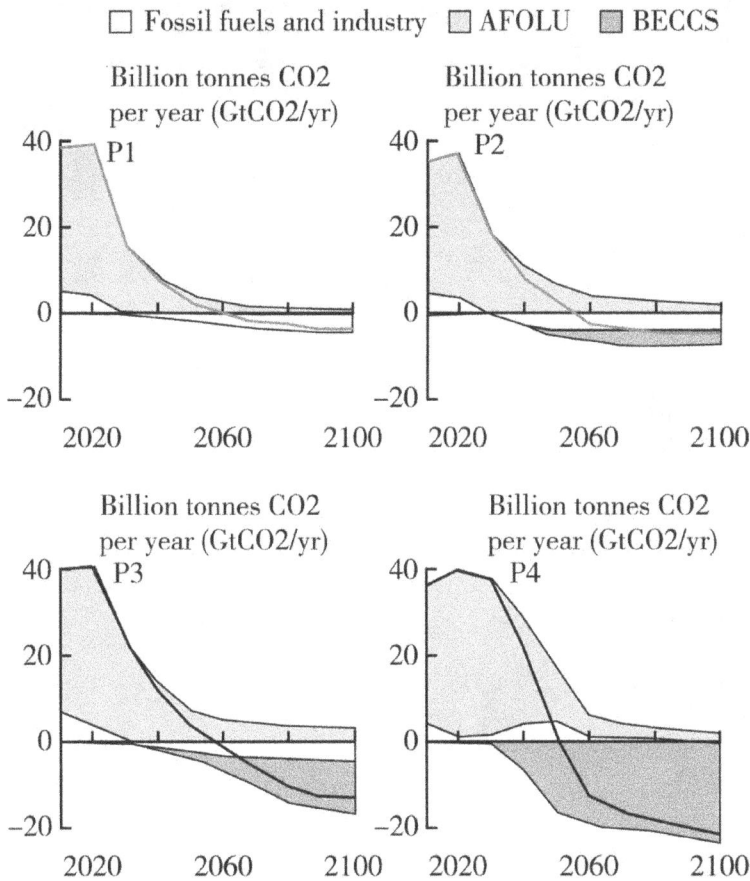

Figure 7. Breakdown of contributions to global net CO$_2$ emissions in four illustrative model pathways
Source: the IPCC Special Report on *Global Warming of 1.5 degrees Celsius* (SR15), 2018.

direct investments. But in the meantime, they may have an adverse impact on actions to tackle climate change. For example, manufacturing and transporting industrial products will produce huge greenhouse gas emissions. When faced with specific requirements for implementing climate change measures, multinational corporations which invest in carbon-intensive industries will protect their investment interests through relevant international investment treaties. These treaties have three main forms: bilateral investment agreements, regional investment agreements, and free trade agreements. Basically, these three types of international investment agreements all aim to promote

Table 3. Breakdown of contributions to global net CO_2 emissions in four illustrative model pathways

Global indicators	P1	P2	P3	P4
Pathways classification	No or limited overshoot	No or limited overshoot	No or limited overshoot	Higher overshoot
CO_2 emission change in 2050 (% rel to 2010)	−93	−95	−91	−97
Kyoto-GHG emissions* in 2050 (% rel to 2010)	−82	−89	−78	−80
Final energy demand** in 2050 (% rel to 2010)	−32	2	21	44
Renewable share in electricity in 2050 (%)	77	81	63	70
Primary energy from coal in 2050 (% rel to 2010)	−97	−77	−73	−97
from oil in 2050 (% rel to 2010)	−87	−50	−81	−32
from gas in 2050 (% rel to 2010)	−74	−53	21	−48
from nuclear in 2050 (% rel to 2010)	150	98	501	468
from biomass in 2050 (% rel to 2010)	−16	49	121	418
from non-biomass renewables in 2030 (% rel to 2010)	833	1327	878	1137
Cumulative CCS until 2100 ($GtCO_2$)	0	348	687	1218
Cumulative BECCS until 2100 ($GtCO_2$)	0	151	414	1191
Land area of bioenergy crops in 2050 (million km^2)	0.2	0.9	2.8	7.2
Agricultural CH_4 emissions in 2050 (% rel to 2010)	−33	−69	−23	2
Agricultural N_2O emissions in 2050 (% rel to 2010)	6	−26	0	39

Note: * Kyoto-gas emissions are based on IPCC Second Assessment Report GWP-100

** Changes in energy demand are associated with improvements in energy efficiency and behavior change

Source: The IPCC Special Report on *Global Warming of 1.5 degrees Celsius* (SR15), 2018.

and protect the interests of foreign investors, and to restrict or prohibit host governments from imposing regulatory measures on investors whose interests will be thus adversely affected and damaged. Therefore, they may ignore the sovereign power of host countries to regulate economy and protect the environment. Biased in favor of the interests of investors, these agreements are contradictory with international agreements and treaties on climate change, such as the United Nations Framework Convention on Climate Change and the Kyoto Protocol. Therefore, countries around the world have come to realize that implementing measures to tackle climate change will require international investors to bear new responsibilities and burdens, and whether these responsibilities and burdens will in turn prevent governments from taking necessary measures to reduce emissions.

34. Why Is It Still Essential to Make Relentless Efforts to Adapt to Climate Change Even If the 1.5 Degrees Celsius Target Is Achieved in the Future?

Climate change adaptation refers to the process of adapting to current or expected climate change and its effects. Although climate change is a global issue, it has different impacts on different parts of the world, which means that natural ecosystems and socioeconomic systems often respond differently and specifically to local climate change, and people in different regions need to adapt in different ways. If global warming increases from 1.2 degrees Celsius to 2 degrees Celsius or higher above pre-industrial levels, climate change adaptation will be more urgently needed. Of course, adapting to global warming of 1.5 degrees Celsius is less demanding than adapting to global warming of 2 degrees Celsius. However, even in the case of a warming of 1.5 degrees Celsius, impacts of previous or current greenhouse gas emissions will last for decades, centuries, or even longer, even if we take mitigation measures immediately to reduce greenhouse gas emissions and endeavor to achieve net-zero emissions. For example, even if we succeed in limiting global warming to 1.5 degrees Celsius by the end of the 21st century, the instability of the Antarctic Ice Sheet and/or the irreversible loss of the Greenland Ice Sheet will continue to cause sea level to rise by several meters in the coming hundreds to thousands of years. Therefore, it is still essential to take relentless efforts to adapt to climate change.

Many adaptation practices, including strengthening disaster early warning and forecasting, investing in flood defenses, building seawalls or restoring mangroves, ecosystem-based adaptation, strengthening biodiversity management, developing sustainable aquaculture, helping people move out of high-risk areas, choosing drought-tolerant crops to avoid yield losses, establishing sustainable water management systems, and building capacity to better respond to impacts of climate change, strengthening financing mechanisms, such as providing different types of insurance and enhancing the popularization of knowledge, are lessening adverse impacts of climate change.

As adaptation is still in its infancy in many regions, adaptability of vulnerable populations remains worrisome. Successful adaptation requires strong support from governments at national and local levels because they can play an important role in coordination, planning, policy priorities setting, resources allocation, and support giving. Given the fact that climate risks are very different in different regions, requirements for climate change adaptation and measures to reduce climate risks in each region must be very tailor-made.

Successful adaptation will minimize adverse impacts of climate change. For example, farmers have chosen drought-tolerant crops to cope with increasingly frequent heat waves and built seawalls to stop flooding caused by climate change-induced sea level rise. In some cases, impacts of climate change may lead to major system-wide changes. For example, a changed climate requires local citizens to choose a new agricultural system and cities to improve urban planning for better flood management. These actions need institutional and organizational changes as well as greater financial support. Adaptation is an iterative process that requires refinements and revisions of adaptation strategies based on continuous assessments on specific adaptation actions. It is also necessary to fully consider the trade-off between particular adaptation options. For example, upstream rainwater collection may reduce downstream water availability; though desalination plants may improve current water supply, they will have huge energy demand over time. So, adaptation measures need to be fully evaluated.

35. How Should Coastal Communities and Populations Better Adapt to Climate Change?

Climate change has caused continuous global mean sea level rise, and this trend will last for centuries. Between 1901 and 1990, the global mean sea level rose by 1.5 mm per year, and it soared to 3.6 mm between 2005 and 2015. By the end of this century, the global sea level may rise by 0.26–0.82 meters, or even higher. By 2300, it may rise by more than three meters, if greenhouse gas emissions and the Antarctic Ice Sheet react to such a degree. Without ambitious adaptation measures, the combined effects of multiple disasters, such as coastal storms and ultra-high tides, will significantly increase the frequency and severity of floods in low-lying coastal areas.

Coastal areas are the most densely populated and developed areas, and house numerous assets and important resources. Many of these areas have adopted a range of measures to address coastal hazards exacerbated by sea level rise, such as coastal floods, coastal erosion, and salinization caused by extreme events, such as storm surges and tropical cyclones. However, many coastal communities are still unable to fully adapt to the current extreme sea levels.

Sea-level rise and adaptation measures may vary in different regions. So, adaptation measures in coastal areas should be tailor-made. Take some hard measures for coastal protection, for example. Though building dikes and sea-walls can effectively help reduce risks caused by a sea level rise of two meters or even higher, they cannot be built endlessly. In densely populated low-lying coastal areas, such as many coastal cities and small islands, the benefits of such protection measures tend to outweigh the costs. However, poverty-stricken areas can hardly bear the costs of hard protection measures. Healthy coastal ecosystems, such as mangroves, seagrass beds, or coral reefs, can be maintained through soft measures for coastal protection. In coastal areas where risks are very high and cannot be effectively reduced, retreat from coastlines is the only way to eliminate risks. Millions of people living on low-lying islands, including residents in small island developing states, some densely populated but less developed deltas, coastal rural villages and towns, and the Arctic which has seen sea ice melting and unprecedented climate change, all face serious situations. Therefore, coastal countries, cities, and regions need to take more urgent and tailor-made adaptation actions.

POLICY ACTIONS
FOR ACHIEVING CARBON PEAKING
AND CARBON NEUTRALITY

Global climate change has constantly rung the alarm bell for human society. Severely challenged by climate change, the international climate process is making slow progress in the midst of hardships. Achieving carbon peaking and carbon neutrality requires stronger policy actions, which means that we need a comprehensive and profound socioeconomic transformation. Starting with current situations and trends in global and China's carbon emissions, this chapter briefly reviews international climate governance and China's contributions, and focuses on challenges, opportunities, and transformation pathways of achieving carbon peaking and carbon neutrality in terms of technology, policy actions, and other aspects.

Section I: Current Situations and Trends in Global and China's Carbon Emissions

36. What Are Current Situations of Global Greenhouse Gas Emissions?

According to the latest figures reported by the PBL Netherlands Environmental Assessment Agency, total global greenhouse gas emissions have increased

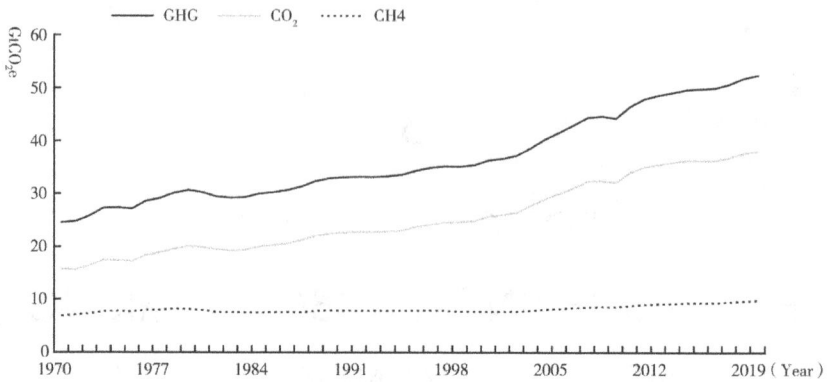

Figure 8. Total global GHG emissions and major GHG emissions, 1970–2019
Note: Total GHG emissions exclude those from land use change.
Data source: Netherlands Environmental Assessment Agency, *Trends in Global CO_2 and Total Greenhouse Gas Emissions; 2020 Report.*

by an average of 1.4 percent annually since 2010. A record high was reached in 2019, with total emissions excluding those from land use change reaching 52.4 gigatonnes of carbon dioxide equivalent, up by 44 percent and 59 percent from the 2000 level and the 1990 level respectively, and global greenhouse gas emissions per capita reaching 6.8 tonnes of carbon dioxide equivalent. Including emissions from land use change (5.5 gigatonnes of carbon dioxide equivalent), the 2019 global greenhouse gas emissions amounted to 59.1 gigatonnes of carbon dioxide equivalent (Figure 8).

Between 2010 and 2019, CO_2 emissions (from fossil fuels and carbonates) accounted for about 72.6 percent of total global greenhouse gas emissions (excluding those from land use change), making it the largest source of greenhouse gas emissions. Emissions from methane (CH_4) and nitrous oxide (N_2O) accounted for about 19 percent and 5.5 percent respectively, and the remaining 2.9 percent of emissions were from fluorinated gases, such as hydrofluorocarbons (HFCs), perfluorocarbons (PFCs), and sulfur hexafluoride (SF_6) (Figure 9).

According to the International Energy Agency (IEA), global carbon dioxide emissions from fossil fuels reached a record of 38 gigatonnes in 2019. Emissions from coal, oil, and natural gas accounted for 43.8 percent, 34.6 percent, and 21.6 percent of total carbon dioxide emissions from fossil fuels respectively, and coal combustion emits almost twice as much carbon dioxide per unit of energy as does the combustion of natural gas. Electricity and heat, transport, and industry are the sectors with largest carbon dioxide emissions in the world,

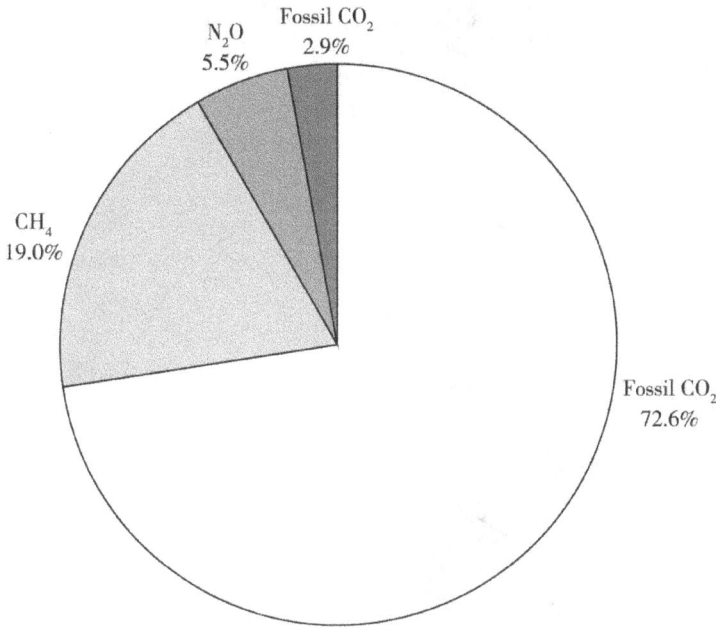

Figure 9. Sources of GHG emissions (excluding those from land-use change), 2010–2019
Source: UNEP, *Emissions Gap Report 2020*.

accounting for about 85 percent of total carbon dioxide emissions from fossil fuels (Figure 10).

Between 2010 and 2019, the world's top six GHG emitting countries (regions) contributed to 62.5 percent of total greenhouse gas emissions (excluding those from land use change). China contributed to more than a quarter of total greenhouse gas emissions, with emissions per capita about 40 percent higher than the global average. The United States contributed to 13 percent of global greenhouse gas emissions, with the emissions per capita three times higher than the global average. 27 European Union countries and the United Kingdom contributed to 8.6 percent of global greenhouse gas emissions, with emissions per capita 25 percent higher than the global average and an annual average decline of about 1.5 percent over the past decade. India contributed to 7.1 percent of global greenhouse gas emissions, with emissions per capita about 60 percent below the global average. Russia and Japan contributed 4.8 percent and 2.8 percent respectively toward global greenhouse gas emissions respectively, with emissions per capita reaching 17.4 and 10.7 tonnes of carbon dioxide equivalent respectively in 2019. See Figures 11 and 12.

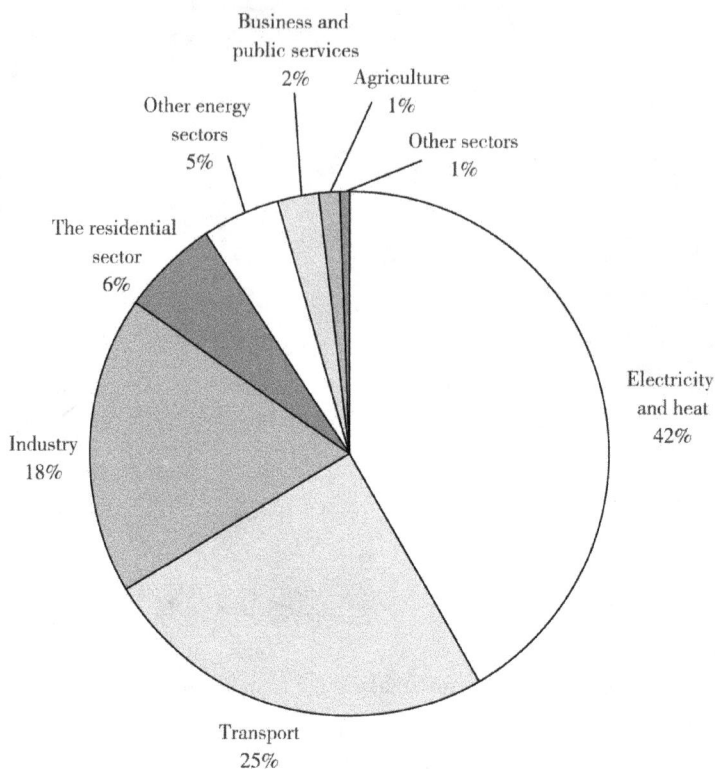

Figure 10. Sources of global fossil CO_2 emissions (by sector, 2018)
Data source: IEA, https://www.iea.org/subscribe-to-data-services/co2-emissions-statistics.

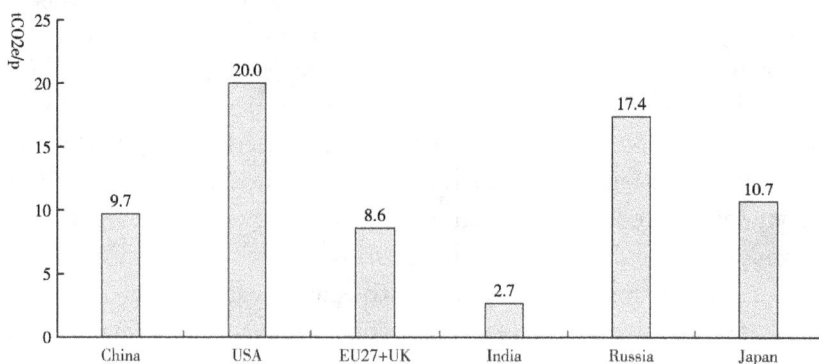

Figure 11. GHG emissions per capita of major global emitters, 2019
Source of data: UNEP, *Emissions Gap Report 2020*.

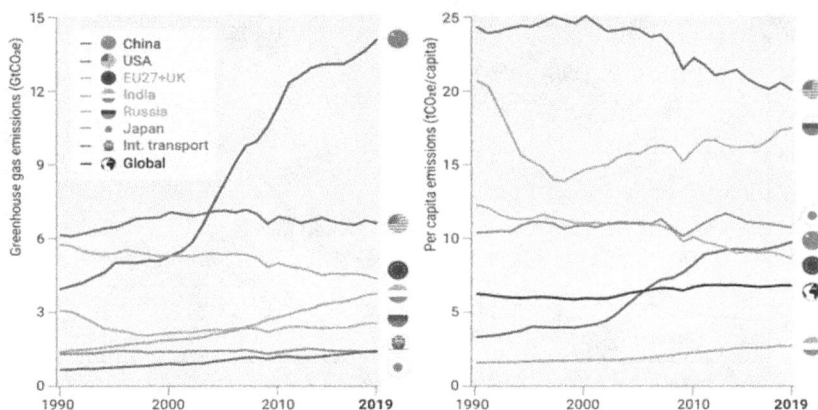

Figure 12. Absolute GHG emissions of the top six emitters (excluding LUC emissions) and international transport (left) and per capita emissions of the top six emitters and the global average (right)
Source of data: UNEP, *Emissions Gap Report 2020*.

37. What Are Current Situations of China's Greenhouse Gas Emissions?

China is one of the countries with the largest total greenhouse gas emissions in the world. In 2019, China's greenhouse gas emissions reached 14 gigatonnes of carbon dioxide equivalent, with emissions per capita of about 9.7 tonnes of carbon dioxide equivalent, accounting for about 27 percent of total global greenhouse gas emissions (excluding those from land use change). Between 2010 and 2019, China's total greenhouse gas emissions increased by an average of about 2.3 percent annually over the past decade, higher than the global average. Since 2010, China's total greenhouse gas emissions have increased by about 24 percent, and CO_2 emissions have increased by 26 percent (Figure 13)

In 2019, China's carbon dioxide emissions accounted for 82.6 percent of the total greenhouse gas emissions, about 10 percent higher than the global average. In addition, 11.6 percent of greenhouse gas emissions were from methane, and about 3 percent and 2.8 percent from nitrous oxide and fluorinated gases (Figure 14).

According to the International Energy Agency, carbon dioxide emissions from coal combustion in China reached 7.61 gigatonnes in 2018, accounting for 80 percent of China's total carbon dioxide emissions from fossil fuels (Figure 15). In addition, 14 percent of carbon dioxide emissions were from oil and

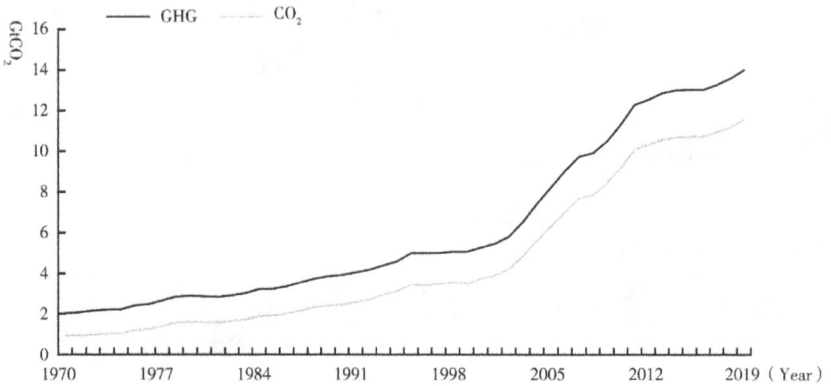

Figure 13. China's GHG and CO_2 emissions, 1970–2019
Data source: Netherlands Environmental Assessment Agency, *Trends in Global CO_2 and Total Greenhouse Gas Emissions; 2020 Report.*

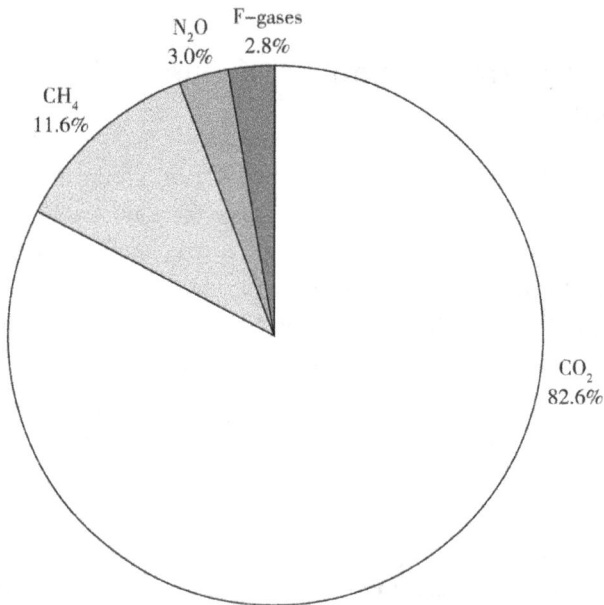

Figure 14. Sources of China's GHG emissions, 2019
Data source: Netherlands Environmental Assessment Agency, *Trends in Global CO2 and Total Greenhouse Gas Emissions; 2020 Report.*

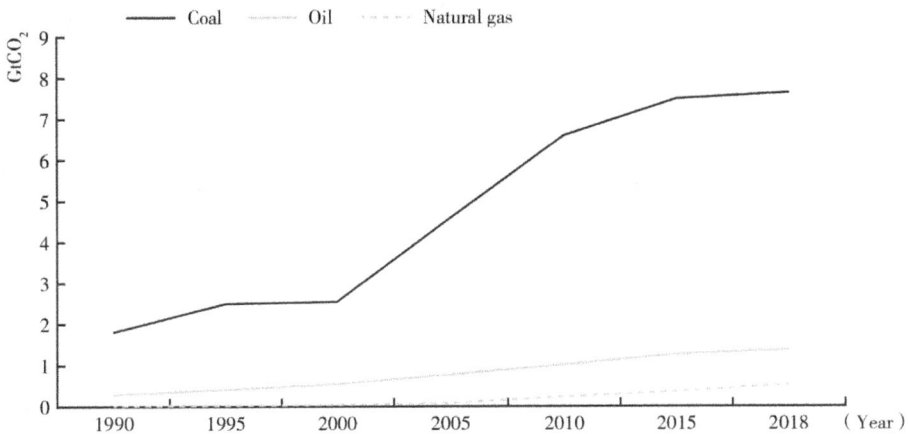

Figure 15. China's CO_2 emissions (by energy source)
Data source: IEA, https://www.iea.org/subscribe-to-data-services/world-energy-balances-and-statistics.

6 percent from natural gas. Coal consumption has a great impact on the trend in China's greenhouse gas emissions. In 2018, coal accounted for 56 percent of China's primary energy supply, renewable energy and nuclear energy 17 percent, oil 19 percent, and natural gas 8 percent. Therefore, fossil fuels accounted for about 83 percent of China's primary energy supply. Emissions from electricity and heat accounted for more than half of China's total carbon dioxide emissions from fossil fuels. Industry, transport, housing, and other sectors are also major sectors of carbon dioxide emissions (Figure 16).

38. What Are the Main Factors Affecting Carbon Emissions?

Carbon emissions, namely carbon dioxide (CO_2) emissions, refers to the total carbon dioxide emissions in a country or region over a certain period (usually a year or an accounting period). Carbon dioxide is emitted in the processes of fossil fuel consumption and industrial production. The concepts involving carbon emissions include national total carbon emissions, national cumulative carbon emissions, carbon emissions per capita, cumulative carbon emissions per capita, the rate of cumulative carbon emissions per capita, etc. A country's carbon emissions per capita are mainly affected by the following socioeconomic factors:

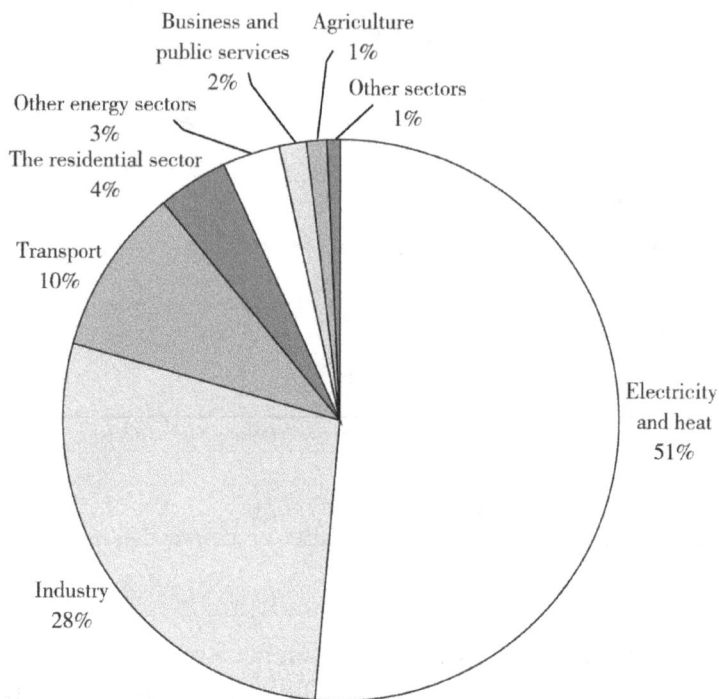

Figure 16. Sources of China's fossil CO_2 emissions (by sector, 2018)
Data source: IEA: https://www.iea.org/subscribe-to-data-services/co2-emissions-statistics.

Stages of economic development. Economic development manifests itself in industrial structure, income per capita and the level of urbanization. Industrial expansion has caused huge fossil CO_2 emissions. Developing the tertiary industry and upgrading industrial structure can contribute to carbon dioxide emissions reduction. A higher income per capita will help increase the capability and willingness of a country's residents to pay for environmental goods. Developed countries are in the post-industrial era and have completed urbanization. Carbon emissions of these countries are mainly driven by a consumption-oriented society. On the contrary, developing countries, such as China, are still in the accumulative phase of economic development. Carbon emissions of these countries are mainly driven by investments in production and infrastructure.

Endowment of energy resources. Carbon emissions mainly comes from the use of fossil fuels. The carbon emissions coefficients of coal, oil, and natural gas are descendent. Green plants are carbon-neutral. Renewable energy, such as solar energy, wind energy, and hydropower, as well as nuclear energy, belong to zero-carbon energy. The endowment of a country's energy resources has a significant impact on carbon emissions. Abundant low-carbon resources are crucial for carbon emissions reduction. Increasing the share of clean energy and promoting the transformation of energy structure will help reduce carbon intensity.

Technological factors. Technological progress can help slow down or even reduce carbon dioxide emissions in the form of enhanced energy efficiency, management efficiency, and carbon capture and storage.

Consumption patterns. Energy consumption and its emissions are fundamentally driven by consumption activities of the whole society. Energy consumption patterns and carbon emissions of residents in different countries may differ significantly because development levels, natural conditions, and lifestyles of each country are quite different. Consumption patterns, behaviors, and habits have a significant impact on carbon emissions. For example, the emissions per capita of the United States are more than twice those of European Union countries.

In addition, demographic changes, environmental policies, and international environment also have a significant impact on a country's carbon emissions.

39. What Impacts Has International Trade Exerted on Carbon Emissions?

Production and consumption activities are the main sources of carbon emissions, and trade activities are a major cause of more carbon emissions. In the context of economic globalization, production and consumption behaviors have been crossing borders due to the development of international trade, thus transferring carbon emissions from one region to another and making the allocation of the responsibility for carbon emissions reduction more complex. Developed countries that have opened their economies tend to develop industries with higher added value, and rely on imports with high-energy consumption and high emissions from developing countries. In essence, developed countries have transferred carbon emissions to developing countries through outsourcing of production end of industrial chains. The issue "embodied carbon emissions" in international trade has attracted much attention in international climate governance because it involves international allocation of the responsibility for carbon emissions reduction.

International trade may cause carbon leakage, which occurs when there is an increase in greenhouse gas emissions in a country which do not assume the responsibility for carbon emissions reduction as a result of an emissions reduction by a second country. In reality, developed countries have transferred carbon-intensive industries to developing countries, and imported low-value-added products or semi-finished products from developing countries, thus reducing carbon emissions and fulfilling their target of carbon emissions reduction. But in the process, developing countries has been suffering from the increase in total carbon emissions. The main channels of carbon leakage include international trade of energy products, international transfer of energy-intensive industries, and international trade of carbon-intensive products.

In allocating the responsibility of carbon emissions in international trade, the international community has been divided into two blocs. One bloc holds the principle that producers should be responsible, and the other bloc holds the principle that consumers should be responsible. The former principle is also known as the principle that polluters should be responsible, which requires polluters to pay for consequences caused by themselves. At present, OECD countries basically formulate environmental policies on the basis of this principle, and the IPCC's data on international carbon emissions is also calculated on the basis of this principle and territorial responsibility. However, calculating embodied carbon emissions in international trade based on this principle may have a negative impact on the effectiveness of climate change agreements and cause carbon leakage, which cannot be solved anyway. To offset the shortcomings of the principle that polluters should be responsible, the other principle that consumers should be responsible, which also aims for allocating the responsibility of carbon emissions reduction in international trade, requires consumers to be responsible for any impacts that production has on the whole ecology and all carbon emissions. Reallocating and recalculating carbon emissions in terms of consumption will effectively help prevent and reduce carbon leakage and promote the diffusion and spillover of low-carbon technologies.

The principle of common but differentiated responsibilities under the United Nations Framework Convention on Climate Change and the Kyoto Protocol requires that developed countries responsible for historical carbon emissions should take the lead in undertaking the obligation of quantitative carbon emissions reduction. Therefore, developed countries have taken actions to implement stricter environmental regulations on carbon emissions at home. However, this has led to another important dispute in international trade, namely the dispute over the border adjustment tax in the form of carbon tariffs. Based on comparative advantages, developed countries have transferred some carbon-intensive industries to developing countries. In the name of protecting the competitiveness of their export enterprises in international and domestic markets, they have been attempting to impose special carbon dioxide emission tariffs on emission-intensive products imported from countries that do not levy carbon taxes or energy taxes but give substantial energy subsidies to aluminum, steel, cement, and some chemical products. But actually, they intend to weaken the competitiveness of competitors and impose trade protectionism in the name of tackling climate change. If we analyze in detail, we can find that imposing carbon tariffs will not make significant contributions to carbon emissions reduction; instead, it will only make both developed countries and developing countries less competitive in international trade, and cause a significant decline in the welfare of all parties.

China enjoys long-standing comparative advantage in labor. Known as the factory of the world, China has become a hub for global manufacturing and produced numerous products which will be consumed by other countries. In 2020, China's total import and export value was 32.16 trillion yuan, with the export value reaching 17.93 trillion yuan. This means that huge carbon emissions from other countries have been transferred to China on the production end. Fast growth of China's carbon emissions is attributed to the expansion of domestic investments, domestic consumption demand, and foreign consumption demand. China is not the only party that should be responsible for embodied carbon emissions in international trade. Consumers in developed countries should also be responsible. Facing some disputes over carbon emissions in international trade, China should carry out in-depth studies and scientific assessments on impacts of international trade on China's carbon emissions, so as to have a greater say in future international climate negotiations.

40. Is Agriculture a Source or Sink of Carbon Emissions?

Affected by natural and human activities, agriculture, forestry and other land use (AFOLU) has become both an important carbon dioxide (CO_2) sink (such as afforestation and soil carbon sequestration management) and an important source of carbon dioxide (CO_2), methane (CH_4), and nitrous oxide (N_2O) emissions (such as deforestation, peatland drainage). Agricultural activities are a major source of non-CO_2 emissions, such as CH_4 emissions from livestock and poultry farming and rice cultivation, and N_2O emissions from manure management, agricultural soils, and biomass combustion. Between 2007 and

2016, global AFOLU greenhouse gas emissions accounted for 22 percent of total anthropogenic greenhouse gas emissions. Technological measures, such as improving farmland water and fertilizer management, improving animal management and grazing management, reducing grassland reclamation, and increasing agroforestry systems, have great potential for CH_4 and N_2O emissions reduction and carbon sequestration. Afforestation, forest management, and prevention of deforestation and forest degradation are important measures to increase carbon storage of forest ecosystems. Low-cost emissions reduction technologies in agriculture and forestry have great potential for carbon emissions reduction and sequestration, and can greatly help reduce environmental pollution, such as air pollution, water pollution, and soil pollution.

Biological carbon sequestration is a process of controlling carbon fluxes to improve carbon absorption and storage of ecosystems through photosynthesis of plants. It is the cheapest way to fix carbon dioxide in the atmosphere and has the least side effects. Biological carbon sequestration mainly involves three forms of technologies: 1. Protecting existing carbon pools through ecosystem management technologies to strengthen the management of agriculture and forestry and to maintain the long-term carbon sequestration capacity of ecosystems; 2. Expanding carbon pools to improve carbon sequestration. Land use change as well as seed selection, plant breeding, and planting technologies can help improve the productivity of plants and carbon sequestration; 3. Producing biological products in a sustainable way, such as replacing fossil fuels with bioenergy.

At present, the commonly used measures to reduce emissions and increase sinks in farming include reducing nitrogen fertilizers, deep root fertilization, water and fertilizer management in paddy fields, application and adding of fertilizers with nitrification inhibitors in paddy fields, and application of decomposed organic fertilizers in paddy fields, application of enhanced efficiency fertilizers in drylands, replacing chemical fertilizers with organic fertilizers, choosing high-yielding and low-emission varieties for breeding, enhancing biodiversity, optimizing planting times, using biochar, choosing conservation tillage, and using straws as organic fertilizer. Measures to reduce emissions and increase sinks in animal husbandry include choosing low-protein diets, improving livestock breeding, choosing feed additives, improving concentrate-to-roughage ratios, improving the quality of roughages, covering liquid manure, manure acidification, using household biogas, choosing biogas engineering, zero grazing, semi-intensive grazing, extensive grazing, etc. Measures to reduce emissions and increase sinks in forestry include artificial afforestation, woodland management, forest harvesting, management of forest disasters and forest products, etc. Measures to reduce emissions from and increase sinks of wetlands include restoration and reconstruction of wetland vegetation, hydrologic restoration of wetlands, improvement of wetland matrix, etc.

41. How Does Land Use Change Affect Global Climate Change?

Land use change is one of the most important factors of global climate change. The development of industrialization and urbanization of human society has caused land use and land cover changes, thus directly changing physical characteristics of land surface and exchanges of energy and substances between land surface and the atmosphere, affecting the energy balance of land surface, and having a great impact on regional climate change. Changes in the types and density of vegetation on land surface and related soil properties will also affect terrestrial carbon storage and fluxes, thus changing atmospheric concentrations of greenhouse gases.

As human activities change vegetation properties on a large scale, the albedo of the Earth's surface will be affected. For example, the albedo of farmlands is very different from that of natural vegetation, such as forests. The albedo of the forest floor is usually lower than that of open lands. Trees in forests have many large leaves, and the incident solar radiation will be reflected and refracted for many times in forest canopies, thus decreasing the albedo of the forest floor. Human-induced aerosols in the atmosphere also affect the albedo of the Earth's surface. It is noteworthy that black carbon above snow-covered lands greatly lowers the albedo of the Earth's surface.

Land use change can also cause some other physical properties of the Earth's surface, such as the ratio of long-wave radiation emitted from the Earth's surface to the atmosphere, soil moisture, and roughness of the Earth's surface, to be different. It can affect the inflow and outflow of energy and water vapor on the Earth's surface through various energy exchanges between the Earth's surface and the atmosphere, thus directly changing atmospheric temperature, humidity, precipitation, and wind speed near the Earth's surface, and having an impact on local and regional climate to some degree. For example, the Amazon rainforest in South America has a significant impact on the temperature and water cycle of the land surface in this region. About half of rainfall over the Amazon Basin derives from evaporation from the Amazon rainforest. Therefore, deforestation of the Amazon rainforest will only lead to changes in the ratios of runoff and evaporation, thus greatly changing regional water cycle. Land use change will also greatly affect regional precipitation and temperature in China. For example, desertification and grassland degradation in northwest China will lead to less precipitation in most areas of China and aggravate droughts in North China and northwest China, thus causing significant temperature increases.

42. How Much Carbon Emissions Have Decreased for the COVID-19 Pandemic? How Much Has the COVID-19 Pandemic Contributed to Mitigating Global Climate Change?

In 2020, human-induced global greenhouse gas emissions decreased by 5.4 percent from the previous year for the COVID-19 pandemic. However, global greenhouse gas emissions in 2021 returned to the 2019 level. Limited reduction in anthropogenic greenhouse gas emissions in 2020 has not significantly changed the concentration of global carbon dioxide emissions, nor has it fundamentally changed the trend in global warming. In 2020, the concentrations of major global greenhouse gases continued to rise, and the global average temperature was about 1.2 degrees Celsius above pre-industrial levels. 2020 was the second warmest year since complete meteorological observation records have been available. The period between 2015 and 2020 were also the warmest six years on record. If 2021 were added, the period between 2015 and 2021 would be the warmest seven years on record.

To explain why reductions in anthropogenic carbon emissions in 2020 have not significantly changed the trend in global warming in vivid language, we still use the volume of water in a swimming pool to represent the volume of carbon dioxide emissions in the atmosphere, and use the changing water level to represent the changing volume of total carbon dioxide emissions in the atmosphere. Even in the absence of man-made carbon emissions, the water level of the pool will fluctuate naturally as rain (representing carbon dioxide emissions from the Earth's natural ecosystems) will raise the water level, and evaporation (representing carbon dioxide absorbed by the Earth's natural ecosystems) will lower the water level. If 10^{15} grams of carbon dioxide in the atmosphere are equivalent to 1 cubic meters of water, each year 110 cubic meters of rainwater will flow into the swimming pool, and evaporation from the surface of the swimming pool will cause basically similar water losses. So, in the natural state, the water level of the swimming pool is basically stable. However, due to seasonal variations of evaporation and precipitation, and the fact that inflow and outflow are not completely synchronous, the water level of the pool is slightly higher in winter and lower in summer, changing within a range of +/−0.5 cm. Since the industrial revolution, however, anthropogenic carbon emissions have increased fast as a faucet had been installed in the swimming pool. Water flowing from the faucet into the swimming pool raises the water level by 11 mm per year, or 0.03 mm per day. Since the industrial revolution, anthropogenic carbon emissions have raised the water level by 64 centimeters. That is to say, it is the 64-centimeter increase in the water level since 1750 that has caused global warming; anthropogenic carbon emissions per day have only raised the water level of the pool by 0.03 millimeters, thus having little impact on global warming. So, a 5–7 percent decrease in anthropogenic carbon emissions by 2020 is equivalent to a 0.002-millimeter decrease in the water level rise per day, thus having a negligible impact on the water level of the pool. In a word, reduction in anthropogenic greenhouse gas emissions in 2020 has not significantly changed the concentration of global carbon dioxide emissions, nor has it fundamentally changed the trend in global warming.

Section II: Global Climate Governance and China's Contributions

43. How Many Stages Does the International Climate Negotiation Process Have?

Since the United Nations Framework Convention on Climate Change was concluded in Rio de Janeiro, Brazil, in 1992, the international community has carried out continuous negotiations on the specification and implementation of the Convention. So far, the international climate negotiation process has undergone several stages: 1995–2005, 2007–2012, 2013–2015, and 2015 onwards, which have seen many landmark achievements, such as the Kyoto Protocol, the Cancun Agreements, and the Paris Agreement.

Between 1995 and 2005, the Kyoto Protocol was negotiated, signed, and entered into force. It was the first executive agreement following the adoption of the Convention. The Convention only stipulated the overall objectives and principles of global cooperative actions. It did not set specific action targets in different stages of global and national actions. Therefore, the Conference of the Parties to the United Nations Framework Convention on Climate Change reached a mandate for negotiations of the Kyoto Protocol in 1995, so as to clarify different targets for global emissions reduction in different stages, tasks that should be undertaken by each country, and modes of international cooperation. As the first executive agreement following the adoption of the Convention, the Kyoto Protocol officially came into force in 2005 after some twists and turns, such as the signing and withdrawal of the United States, and the high asking price of Russia and other countries on emission allowances. As a landmark document, it marks that tasks and targets for each party to reduce emissions under the Convention between 2008 and 2012 have been clarified for the first time. It divided Annex I countries into developed countries and economies in transition, thus shaping the world into the three blocs: developed countries, developing countries, and economies in transition.

Between 2007 and 2012, the international climate regime for 2013–2020 was established after several negotiations. The Bali Road Map was adopted at the 2007 Bali Climate Change Conference in Indonesia, thus initiating the negotiation process of the international climate regime following the Kyoto Protocol and clarifying that the implementation period should span the eight years between 2013 and 2020. According to the mandate of the Bali Road Map, negotiations should have been concluded at the fifteenth session of the Conference of the Parties to the United Nations Framework Convention on Climate Change in 2009. However, the Conference failed to unanimously adopt the Copenhagen Accord, whose main consensuses were written into the Cancun Agreements at the Cancun Climate Change Conference in 2010. In the following two years, decisions adopted by the Conference of the Parties to the United Nations Framework Convention on Climate Change gradually clarified responsibilities and action targets for each party to reduce emissions, thus establishing the post-2012 international climate regime. Afterwards, the Copenhagen Accord, the Cancun Agreements and other agreements no longer made the classification of Annex I countries and non-Annex I countries and cancelled the identification of economies in transition for the eastward expansion of the European Union.

Between 2011 and 2015, the Paris Agreement was concluded after rounds of negotiations, basically establishing the post-2020 international climate regime. In 2011, the seventeenth session of the Conference of the Parties to the United Nations Framework Convention on Climate Change in Durban, South Africa, mandated the launching of the Durban Platform for negotiations of a post-2020 international climate regime. According to the bottom-up action logic established by the Obama administration during negotiations of the Copenhagen Accord, the Paris Agreement in 2015 emphasized the expression of nationally determined contributions that is legally consistent and does not distinguish between countries in the Northern Hemisphere and countries in the Southern Hemisphere so that the differences in self-positioning of countries could only be noticed from the differences in contributions. Henceforth, a global climate governance paradigm under which all countries take joint actions to tackle climate change has been established.

From 2016 onwards, a series of negotiations have been launched on the specification and implementation of rules in the Paris Agreement. In the meantime, the process of international climate governance has once again witnessed the negative impacts caused by regime changes of the United States and Brazil, which have caused a lot of hurdles. In 2018, the twenty-fourth session of the Conference of the Parties to the United Nations Framework Convention on Climate Change in Katowice, Poland, reached basic consensuses on mechanisms and rules involving nationally determined contributions, mitigation, adaptation, finance, technology, capacity-building, transparency, and the global stocktake set out in the Paris Agreement, and made further arrangements for implementing the Paris Agreement and enhancing global actions to address climate change. In 2021, Parties to the Convention completed negotiations on the implementation of rules in the Paris Agreement and adopted the Glasgow Climate Pact.

44. How Has the Basic Landscape of International Climate Negotiations Evolved? What Are the Attitudes of Major Interest Groups in Response to Climate Change?

The basic landscape of global actions on climate change has evolved from the two blocs of countries in the Northern Hemisphere and countries in the Southern Hemisphere in the 1980s to the current situation of the North-South interweaving, the South-North integration, the intra-North divide, and the North-South continuous spectrum. The North-South interweaving means that members of the two blocs of countries in the Northern Hemisphere and countries in the Southern Hemisphere have common interests in geopolitics, economic relations, and climate protection. The South-North integration includes three aspects: some countries in the Southern Hemisphere have become members of the club of developed countries; some countries in the Southern Hemisphere and countries in the Northern Hemisphere have common or similar pursuits of interests; some countries in the Southern Hemisphere still belong to the original bloc but differ from less developed countries, though they have grown into emerging economies and are different from pure countries in the Southern Hemisphere. The intra-North divide means that groups with different pursuits of interests emerge within the bloc of countries in the Northern Hemisphere. The most typical are the Umbrella Group and the European Union, which are sub-divided. Take Poland and Romania, two original economies in transition that later joined the European Union, for example. They hold different stances on climate policies with the original 15 EU countries. It is noteworthy that now countries in the Northern Hemisphere have less control over the global economy, while the status of emerging economies has been dramatically elevated, and the status of less developed countries has remained stable. When it comes to the North-South continuous spectrum, some typical countries and regions will pop into our minds. They can be divided into two blocs, three tiers, and five types of economies. Two blocs refer to countries in the Southern Hemisphere and countries in the Northern Hemisphere. Three tiers refer to developed countries, emerging countries, and less developed countries. Five types of economies include developed economies with rapid population growth, developed economies with stable or declining population growth, emerging economies with stable population growth, emerging economies with rapid population growth, and less developed economies characterized by low income. These countries may continue to be sub-divided and join with each other in the future, and this will be a long-standing trend in the very distant future.

Major interest groups in international climate negotiations include the European Union, the Umbrella Group, small island countries, the BASIC countries, etc. The European Union has been actively participating in climate negotiations and taking climate actions as a whole entity. Major participants of the Umbrella Group include the United States and Russia. Climate actions and policies of the United States have been subject to the influence of the ruling party. To cope with the fluctuation and discontinuity of national policies, local governments, cities, and enterprises have been actively taking climate actions. Believing that climate change may be beneficial to its economic development, Russia has been neutral towards global climate governance. Small island countries, which are vulnerable to existential risks caused by climate change-induced sea level rise, have been particularly concerned about climate change and hope to receive financial support. Among the BASIC countries, India, a country with rapid economic and population growth, tends to adopt relatively moderate policies for emissions reduction. Brazil's climate policies have been constantly subject to regime changes of the ruling party. So, this country has to make a trade-off between economic growth and emissions reduction. South Africa has been always making positive responses to push forward climate negotiations and taking ambitious climate actions. Having never been absent from climate negotiations, China has been taking concrete actions to tackle climate change and urged other countries to actively respond to international climate change through international cooperation.

45. What Role Has the Intergovernmental Panel on Climate Change (IPCC) Played in Advancing the International Climate Process?

In response to challenges caused by climate change, the Intergovernmental Panel on Climate Change (IPCC) was established in 1988 by the World Meteorological Organization (WMO) and the United Nations Environment Programme (UNEP) to help people around the world construct a scientific understanding of the current climate change as well as its potential environmental impacts and socioeconomic impacts. The agency has released a series of reports and provided scientific support to push forward some major events in the process of international climate governance. For example, the IPCC made a systematic assessment on the most recent progress of climate change in the First Assessment Report (FAR), thus lying a scientific foundation for global climate governance and urging the 1992 United Nations Conference on Environment and Development to adopt the first framework international document—the United Nations Framework Convention on Climate Change, which aims at controlling greenhouse gas emissions and addressing global warming; the IPCC Second Assessment Report (SAR) released in 1995, paved the way for the 1997 Kyoto Protocol (hereinafter referred to as "the Protocol"); the IPCC assessed the impacts of climate change on different regions in the Third Assessment Report (TAR), thus making the issue of adaptation as equally important as the issue of mitigation; the IPCC comprehensively assessed future trends in climate change under different concentrations of greenhouse gases in the Fourth Assessment Report (AR4) released in 2007, thus lying a scientific foundation for the 2 degrees Celsius target. Although the Copenhagen Accord released in 2009 did not have any legal effect, the 2 degrees Celsius target has been widely accepted by the international community afterwards; the IPCC clarified the fact of global warming and significant impacts of human activities on the Earth's climate system in the Fifth Assessment Report (AR5), thus laying a scientific foundation for the successful conclusion of the Paris Agreement at the Paris Climate Conference; the IPCC emphasized in the Sixth Assessment Report (AR6) that climate change had a profound impact on the development of human society, and today's choices would determine our future and were expected to have an impact on the first global stocktake starting in 2023. In terms of specific content, the IPCC has continuously enhanced its cognition of different scientific issues through multiple assessments, thus laying a scientific foundation for international climate governance; it has also innovated methods and pathways, and proposed some valuable concepts and ways for better implementation.

46. What Is the Main Content of the Paris Agreement? What Is the Significance of the Paris Agreement?

The Paris Agreement, which was adopted at the twenty-first session of the Conference of the Parties to the United Nations Framework Convention on Climate Change in Paris in December 2015, aimed to control global warming mainly caused by carbon emissions from human activities. It was opened for signature at the United Nations Headquarters in New York on the World Earth Day, April 22, 2016, and entered into force on November 4, 2016. According to the official website of the UNFCCC, 190 parties have signed the agreement. Throughout human history, the Paris Agreement has been the third landmark international legal document to tackle climate change following the United Nations Framework Convention on Climate Change in 1992 and the Kyoto Protocol in 1997. It has made arrangements for post-2020 global actions to tackle climate change.

Under the changing international economic and political landscape, the Paris Agreement is a legal document that aims to achieve the objectives of the UNFCCC and has been concluded to establish the post-2020 international climate landscape. It has 29 articles in total, including objectives, mitigation, adaptation, loss and damage, finance, technology, capacity-building, transparency, the global stocktake, etc. Its core objective is to hold the increase in the global average temperature to well below 2 degrees Celsius above pre-industrial levels and pursue efforts to limit the temperature increase to 1.5 degrees Celsius above pre-industrial levels by strengthening the global response to the threat of climate change. It has established an institutional framework that mainly includes the following aspects:

Firstly, the primary responsibility of developed countries in international climate governance has been further affirmed; the distinction between responsibilities and obligations of developed countries and those of developing countries has been maintained; the intensity and breadth of actions taken by developing countries have been enhanced significantly. As mentioned above, due to the adjustment of the global economy and emission landscapes, developed countries hoped to break the boundary between responsibilities of countries in the Northern Hemisphere and those of countries in the Southern Hemisphere. In the name of requiring all countries to indiscriminately assume the responsibility of tackling climate change and form a unified framework for emissions reduction and monitoring, they actually hoped to transfer the responsibility and obligation of tackling climate change to developing countries. Such stance was opposed by developing countries. In response, they showed strong willingness to cooperate and great flexibility. In the end, the Paris Agreement admitted the gap between countries in the Northern Hemisphere and countries in the Southern Hemisphere, indicating the distinction of responsibilities and obligations of Parties to the Convention. It basically foiled developed countries' attempt to require all countries to indiscriminately assume the responsibility of tackling climate change. The principle of common but differentiated responsibilities was reaffirmed and emphasized in different paragraphs of the document, thus laying a sound foundation for developing countries to participate in international climate governance in an equitable and active manner. The intensity and breadth of actions taken by developing countries have also been significantly enhanced.

Secondly, a bottom-up commitment pattern has been adopted to ensure maximum engagement. The Paris Agreement adhered to consensuses reached in the Copenhagen Accord, allowing Parties to the Convention to independently propose their emissions reduction targets and other targets based on their own socioeconomic development. It was precisely because countries were allowed to propose targets based on their own conditions and willingness to act that many Parties to the Convention that had not proposed nationally determined contributions previously were encouraged to propose their own nationally determined contributions, thus ensuring wide engagement in the Paris Agreement. On the other hand, it was because NDCs were independently proposed by all Parties to the Convention that they seemed more feasible.

Thirdly, the pattern of compulsory and voluntary funding has been established, which contributes to expanding funding channels and nurturing a more diversified funding governance mechanism. The Paris Agreement further clarified the responsibility and obligation of developed countries to provide finance while taking into account developing countries' demand to have differentiated financial obligations. Based on solid facts, this agreement respected differences between countries in the Northern Hemisphere and countries in the Southern Hemisphere. Therefore, it enabled all countries, especially developing countries, to have confidence in participating in international financial cooperation. The Paris Agreement also encouraged all Parties to the Convention to provide voluntary financial support to developing countries to help them tackle climate change. These measures were expected to help all Parties to the Convention consolidate existing funding channels and expand more diversified funding governance models based on mutual trust.

Fourthly, a legal form based on the international political reality has been established. The Paris Agreement embodies both restraint and flexibility. The form of a climate agreement can, to some extent, reflect the political will of countries and the level of global environmental awareness. In 1997, the international community reached the Kyoto Protocol, which clearly clarified the use of a more official legal text—"protocol", to implement the Convention; in 2015, the world generally did much better in tackling climate change than in 1997. China, the United States, and other major emitters, which were conservative in the process of international climate governance, have begun to take proactive steps to tackle climate change. It should be noted that all Parties to the Convention have strengthened their willingness to tackle climate change and done better in this regard. Against such a backdrop, the Paris Agreement, if not legally binding, will not meet the requirement that global environmental awareness be increasingly enhanced and go against the logic that all countries take proactive steps. Although the Paris Agreement was not titled with "protocol", its content, structure, and approval procedures fully met the requirements of a legally binding international treaty. After signatories met certain conditions, the Paris Agreement came into force and became an international law that serves to restrict and regulate post-2020 global climate governance actions. Why wasn't the Paris Agreement titled with "protocol"? On one hand, nationally determined contributions of each country were not included in the document but in the registry outside the Paris Agreement. This determines that it is different from protocols in terms of functions and roles. On the other hand, using the term "protocol" simplifies processes of signatories, thus helping Parties to the Convention get quick ratification.

Fifthly, the global stocktake has been established to dynamically update and enhance efforts to reduce emissions. It will be conducted every five years to ensure the efficient implementation of the Paris Agreement and to urge countries to achieve global long-term emissions reduction targets through nationally determined contributions. It serves to supervise and evaluate how countries implement their own nationally determined contributions, to compare the gap between emissions reduction efforts of the international community and the IPCC's 2 degrees Celsius target or even the 1.5 degrees Celsius target, and to urge countries to enhance efforts to achieve emissions reduction targets through nationally determined contributions or propose new nationally determined contributions. The global stocktake together with "a progression beyond the Party's then current nationally determined contribution" stated in Article 4.3 of the Paris Agreement serves as a guarantee that countries will continue to enhance their actions, to review the adequacy of their actions, and to realize objectives set out in the Paris Agreement and the Convention after the proposal of their nationally determined contributions. Aiming to assess nationally determined contributions of each country, the global stocktake should be implemented in an open, facilitative rather than mandatory way. Together with transparency and the compliance mechanism of the Paris Agreement, the global stocktake imposes pressure on countries that fail to implement nationally determined contributions or those who propose too conservative nationally determined contributions, and urges them to enhance their contributions. Compared with previous climate agreements, the global stocktake is an innovation. It can facilitate and encourage countries which have taken robust actions to continuously make full use of their potential to upgrade their actions; it can also give countries which propose conservative contributions the opportunity to update their contributions and to intensify their actions. In a word, it is conducive to forming a dynamically updated and more proactive mode for global emissions reduction and governance.

47. What Are Nationally Determined Contributions (NDCs)?

Nationally determined contributions (NDCs) refer to action targets set by signatories in accordance with their own conditions to participate in international cooperation on climate change for the ultimate end of achieving objectives set out in the Paris Agreement. They include targets for greenhouse gases control, adaptation, and financial and technical support needed to hold the increase in the global average temperature to well below 2 degrees Celsius. Countries were requested to submit their new NDCs or updated NDCs by 2020 and every five years thereafter. However, most countries have delayed their submissions due to the impacts of the COVID-19 pandemic. Nationally determined contributions represent a country's ambition and goal to reduce emissions nationwide.

Around 2015, a total of 193 parties submitted their intended nationally determined contributions (INDCs). After the Paris Agreement was ratified, most countries automatically converted their INDCs to NDCs. Some countries also proposed conditional NDCs, which means that only when they receive financial and technical assistance, will they take actions to achieve higher emissions reduction targets. On June 30, 2015, China submitted to the United Nations Enhanced Actions on Climate Change: China's Intended Nationally Determined Contributions without setting any conditions, in accordance with our national conditions, development stage, the strategy of sustainable development, and our international responsibility. In this document, China pledged to achieve carbon peaking around 2030 and achieve this goal as earlier as possible, to lower carbon dioxide emissions per unit of GDP by 60–65 percent from the 2005 level, to increase the share of non-fossil fuels in primary energy consumption to around 20 percent, and to increase forest stock by around 4.5 billion cubic meters above the 2005 level. To achieve NDCs by 2030 in response to climate change, China also clearly put forward strengthened policies and measures in terms of institutional mechanisms, production methods, consumption patterns, economy, scientific and technological innovation, and international cooperation.

According to the UNFCCC secretariat, as of October 2021, among all 192 Parties to the Paris Agreement, 165 countries had submitted their upgraded nationally determined contributions, which were expected to cover 94.1 percent of the total global emissions in 2019. It was estimated that 52.4 gigatonnes of carbon dioxide equivalent were emitted globally, excluding those from land use, land-use change and forestry (LULUCF). In December 2020, the European Union adopted a new target of 55 percent emissions reduction by 2030 from the 1990 level. After Brexit, the United Kingdom planned to independently propose a new target of 68 percent emissions reduction by 2030 from the 1990 level. In October 2021, China submitted to the UNFCCC secretariat China's Achievements, New Goals and New Measures for Nationally Determined Contributions and China's Mid-Century Long-Term Low Greenhouse Gas Emission Development Strategy, and updated nationally determined contributions as follows: to lower carbon dioxide emissions per unit of GDP by over 65 percent from the 2005 level; to increase the share of non-fossil fuels in primary energy consumption to around 25 percent by 2030; to increase forest stock by 6 billion cubic meters above the 2005 level; to bring the total installed capacity of wind and solar power to over 1,200 GW.

48. How Big Is the Gap Between NDCs of Each Country and the 2 Degrees Celsius/1.5 Degrees Celsius Target?

According to *Emission Gap Report 2020* released by the United Nations Environment Programme in December 2020, nationally determined contributions (NDCs) committed by countries under the Paris Agreement were still seriously insufficient. Despite a brief drop in carbon dioxide emissions for the sake of the COVID-19 pandemic, the global average temperature rise is still likely to exceed more than 3 degrees Celsius by the end of this century, far beyond the goal of holding the increase in global average temperature below 2 degrees Celsius above preindustrial levels and pursuing efforts to limit it to 1.5 degrees Celsius set out in the Paris Agreement

Emission Gap Report 2020 stated that even if unconditional nationally determined contributions of each country could be fully implemented, global greenhouse gas emissions would still reach 56 gigatonnes of carbon dioxide equivalent by 2030. Based on this figure, it is estimated that the global average temperature rise is still likely to reach 3.2 degrees Celsius by the end of this century. To achieve the goal of holding the increase in global average temperature below 2 degrees Celsius above preindustrial levels, the world needs to control total emissions below 41 gigatonnes by 2030. Based on this figure, it is estimated that there will be an emission gap of 15 gigatonnes of carbon dioxide equivalent by the end of this century. Under the scenario of limiting global warming well below 1.5 degrees Celsius, the emission gap will be as high as 32 gigatonnes of carbon dioxide equivalent. To achieve the 2 degrees Celsius target, countries need to nearly triple their overall emissions reduction committed under the Paris Agreement; to achieve the 1.5 degrees Celsius target, they need to at least quintuple their overall emissions reduction.

The green recovery after the COVID-19 pandemic is expected to help reduce greenhouse gas emissions by about 25 percent from the original forecast of the 2030 level (achieved through the implementation of unconditional NDCs committed by countries). Such expectation is far beyond the results achieved when each country fully implements its unconditional NDCs, greatly increasing the possibility that the world will achieve the 2 degrees Celsius target.

Prior to the UN Climate Change Conference in Glasgow in October 2021, many countries had submitted their new and updated NDCs. *The Emissions Gap Report (EGR) 2021: The Heat Is On* released by the UNEP stated that these mitigation commitments fell far short of what was needed to meet the goals of the Paris Agreement; they only slightly narrowed the gap between the emissions required to achieve the goal of limiting global warming by 2030 set out in the Paris Agreement and the emissions countries committed to achieve by 2030; there was a 66 percent of probability that global warming would reach 2.7 degrees Celsius by the end of this century. To limit global warming well below 1.5 degrees Celsius, the world needs to reduce an additional 28 gigatonnes of carbon dioxide equivalent per year in the following eight years; to achieve the 2 degrees Celsius target, the world needs to reduce an additional 13 gigatonnes of carbon dioxide equivalent per year.

49. Which Countries Have Achieved Net Zero Emissions?

Carbon peaking refers to the occasion when a country's total carbon emissions peak in a certain year and decline steadily thereafter. It is difficult to judge whether a country achieves carbon peaking in a certain year or not, which is subject to verification afterwards. Generally speaking, a country achieves carbon peaking only when its total carbon emissions stop increasing in the five years following the peak. There are different interpretations for the term "carbon". In some cases, it only refers to carbon dioxide emitted from fossil fuel combustion, such as the "carbon" in the goal of achieving carbon peaking and carbon neutrality proposed by China under the Paris Agreement. In other cases, it refers to carbon dioxide equivalent converted from the volume of greenhouse gas emissions. The international community discusses the significance of carbon peaking with a view to judging a country's future trends in carbon emissions and exploring pathways for economic and societal low-carbon development. However, such discussion will only count when countries that pledge to achieve carbon peaking have already undergone fast economic growth and achieved high levels of wealth accumulation and social welfare. It does not make any sense even when countries with low human development and low income achieve carbon peaking one day. With low emissions per capita, these countries should be given the right to more emissions. On the other hand, their future development is enshrouded by great uncertainties. Their peak emissions may only last over a certain period, which is subject to socioeconomic development. According to the data on carbon dioxide emissions of many countries and regions between 1750 and 2019, we analyzed the trends in carbon dioxide emissions from countries and regions beyond high-income countries listed by the World Bank and found that 46 countries and regions (most of them were developed countries, and some were developing countries and regions) had achieved carbon peaking by the end of 2019 (Table 4).

50. How Many Countries Have Committed to Achieve Carbon Neutrality?

Supported by the UNFCCC and the UNDP, Chile and the United Kingdom launched the Climate Ambition Alliance, calling on countries to commit to achieve carbon neutrality by 2050. According to the Energy and Climate Intelligence Unit, a United Kingdom non-profit organization, 127 countries (including the European Union) proposed or committed to propose carbon neutrality in different forms, such as laws, legal proposals, and policy documents. Suriname and Bhutan claimed to have achieved carbon neutrality, which was attributed to their extremely large forest area and low industrialization. Nowadays, an increasing number of countries have taken carbon neutrality as an important strategic goal and taken active measures to tackle climate change (Table 5).

51. What Does the European Green Deal Involve?

In December 2019, President of the European Commission, Ursula von der Leyen, launched the European Green Deal soon after she took office. The European Green Deal is the EU's new growth strategy, aiming to transform the EU into a fairer and more prosperous society, with a modern, resource-efficient and competitive economy, with no net emissions of greenhouse gases by mid-century.

To achieve carbon neutrality by 2050, the European Union proposed seven major tasks in energy, industry, transportation, construction, biodiversity, etc. They are listed as follows: 1. To build a clean, affordable, and secure energy supply system; 2. To facilitate the clean and circular transformation of industrial enterprises; 3. To renovate in a resource/energy-efficient way; 4. To accelerate the establishment of a sustainable smart travel system; 5. To establish a fair, healthy, and environmentally-friendly food supply system; 6. To protect and restore ecosystems and biodiversity; 7. To implement the zero-pollution development strategy for non-toxic environment, including zero-pollution actions for air, water, and soil, and sustainable management of chemicals.

The European Union has also clarified a series of policies to foster green investment and financing, green finance, and green technologies and talents.

Table 4. The years when countries and regions had achieved carbon peaking by 2019 and the maximum emissions

Year	Country	The maximum emissions (10,000 tonnes)	Year	Country	The maximum emissions (10,000 tonnes)
1969	Antigua and Barbuda	126	2003	Finland	7266
1970	Sweden	9229	2004	Seychelles	74
1971	Britain	66039	2005	Spain	36949
1973	Brunei	997	2005	Italy	50001
1973	Switzerland	4620	2005	America	613055
1974	Luxembourg	1443	2005	Austria	7919
1977	Bahamas	971	2005	Ireland	4816
1978	The Czech Republic	18749	2007	Greece	11459
1979	Belgium	13979	2007	Norway	4623
1979	France	53028	2007	Canada	59422
1979	Germany	111788	2007	Croatia	2484
1979	Netherlands	18701	2007	Taiwan, China	27373
1984	Hungary	9069	2008	Barbados	161
1987	Poland	46373	2008	Cyprus	871
1989	Romania	21360	2008	New Zealand	3759
1989	The Bermuda Triangle	78	2008	Iceland	382
1990	Estonia	3691	2008	Slovenia	1822
1990	Latvia	1950	2009	Singapore	9010
1990	Slovakia	6163	2010	Trinidad and Tobago	4696
1991	Lithuania	3785	2012	Israel	7478
1996	Denmark	7483	2012	Uruguay	859
2002	Portugal	6956	2013	Japan	131507
2003	Malta	298	2014	Hong Kong, China	4549s

Source: public statistics from Our World in Data, https://ourworldindata.org/grapher/annual-co2-emissions-per-country?tab=chart.

Click https://datahelpdesk.worldbank.org/knowledgebase/articles/906519-world-bank-country-and-lending-groups for information about how the World Bank classifies high-come countries.

Note: The maximum emissions refer to the highest values of carbon dioxide emissions (excluding those from land use change) in the year when each country achieved carbon peaking.

These policies include increasing public funds in green investment, facilitating green financing channels in the private sector, advocating fair transformation, using green budget tools to enhance the priority of green projects in public investment, accelerating reforms of energy tax and other types of tax, promoting research and development of green technologies, and accelerating the construction of digital infrastructure.

The European Green New Deal is also committed to being a political leader in global climate actions by implementing powerful green diplomacy and facilitating the world to improve policy tools for better tackling climate change, including establishing a global carbon market, promoting the EU green standards, and improving global sustainable financing platforms. It is noteworthy that the European Union emphasizes the need to improve the status of climate change in trade policies. Its establishment of the Carbon Border Adjustment Mechanism (CBAM) to enhance the access standards of imports in specific industries, such as food, chemicals, and materials, has attracted great attention from the international community.

In March 2020, the European Union officially submitted "the long-term low greenhouse gas emission development strategy" to the UNFCCC secretariat. After the outbreak of the COVID-19 pandemic, the European Union reiterated its commitment to implement the European Green New Deal to facilitate green recovery. On September 17, 2020, the European Commission released the 2030 Climate Target Plan. It stated that by 2030, greenhouse gas emissions would decline by at least 55 percent from the 1990 level, significantly higher than the previous 40 percent reduction target. Despite different development levels and specific national conditions within the European Union, each member had its own pursuit for emissions reduction targets. For example, Eastern European countries, represented by Poland, which are highly dependent on fossil fuels, called on the EU not to formulate generalized climate policies and demanded financial assistance. On December 11, 2020, with relentless efforts, leaders from the European Union reached an agreement which stated each member, including Poland, had agreed to raise the EU's 2030 target of reducing carbon emissions by the current 40–55 percent. In the meantime, the European Union also made significant progress in climate laws, making the goal of achieving net zero emissions of greenhouse gases by 2050 legally binding on EU institutions and member states.

Britain is the first major economy to legally commit to achieve net zero emissions by 2050. It has rich theoretical and practical experience in both reducing emissions and promoting economic growth. In June 2020, Prime

Table 5. Carbon neutrality proposed by major countries in the world (including the European Union)

Country/Party to the convention	Nature of commitment	Committed time frame for carbon neutrality
Suriname	—	Achieved
Bhutan	—	Achieved
Denmark	In law	2050
France		2050
Hungary		2050
New Zealand		2050
Sweden		2045
Britain		2050
Canada	Proposed	2050
Chile		2050
The European Union		2050
Spain		2050
South Korea		2050
Fiji		2050
Finland	In policy documents	2035
Austria		2040
Iceland		2040
Japan		2050
Germany		2050
Switzerland		2050
Norway		2050
Ireland		2050
South Africa		2050
Portugal		2050
Costa Rica		2050
Slovenia		2050
Marshall Islands		2050
America		2050 (Biden's campaign promise)
China		2060
Singapore		The second half of the 21st century
Dozens of other countries	In discussion	2050

Source: Energy & Climate Intelligence Unit, https://eciu.net/netzerotracker.

Climate Ambition Alliance: Net Zero 2050, https://climateaction.unfccc.int/.

Minister Boris Johnson announced a New Deal to stimulate the economy with the slogan of "Build, Build, Build." In November, he announced The Ten Point Plan for a Green Industrial Revolution, worth of £12 billion, of which £1 billion would be invested in carbon capture and storage. He promised that Britain would strive to be at the global forefront of carbon capture, utilization and storage, and to establish four carbon capture and storage centers and clusters by 2030, which could support up to 50,000 jobs.

The Russian-Ukrainian war that began in early 2022 has caused volatility in global energy markets, bringing about significant implications and uncertainties for global energy transition, especially the energy transition and climate policies of Europe. On one hand, some European countries, which are highly dependent on energy imports from Russia, have temporarily slowed down their plans to close coal power plants, or even restarted the use of coal power and nuclear power, so as to cope with the shortage of energy supply and soaring energy prices. This will adversely affect the realization of carbon neutrality to some degree. On the other hand, the European Union has been actively seeking the diversification of energy import channels, vigorously saving energy, and accelerating the development of renewable energy. In the long run, these actions can fundamentally help EU member states to free from high dependence on Russia energy imports.

52. How Did the Biden Administration Implement Its Climate Policies?

During the election campaign, Biden fought with Trump over climate change. He campaigned on a platform of Clean Energy Revolution and Environmental Justice, supporting the Select Committee on the Climate Crisis to release the document Solving the Climate Crisis: The Congressional Action Plan for a Clean Energy Economy and a Healthy, Resilient, and Just America. On January 20, 2021, Biden was sworn in as the 46th President of the United States. On Jan 27, 2021, he signed a new executive order to tackle climate change, saying that having waited too long, it was time for the United States to take actions to deal with the climate crisis.

The first climate policy of the Biden administration is to coordinate climate actions at home and abroad. Biden appointed Gina McCarthy, a former EPA administrator, as domestic climate policy coordinator to lead the White House Office of Domestic Climate Policy. He also appointed former Secretary of State of the United States, John Forbes Kerry, as Special Presidential Envoy for Climate to coordinate foreign climate policies and to integrate climate change into every aspect of U.S. domestic and foreign policies. The second climate policy of the Biden administration is to abandon Trump's isolationism to reshape America's leadership in the international climate governance system. Biden signed a series of executive orders on his first day in office to bring the United States back into the Paris Agreement. On February 19, 2021, the United States formally rejoined the Paris Agreement. On April 22, 2021 (the World Earth Day), Biden hosted the Virtual Leaders Summit on Climate, announcing a 50–52 percent reduction in national greenhouse gas emissions by 2030 from the 2005 level and net zero emissions by 2050. This marks the resumption of the U.S. government's participation in international climate governance. The third climate policy of the Biden administration is to vigorously develop clean energy and reduce methane emissions. Committing to use 100 percent clean electricity and 100 percent clean energy vehicles by 2035, to achieve net zero emissions from trucks and buses by 2040, to freeze oil and gas leases on United States federal land, and to reduce methane emissions from oil and gas exploration activities, is the most effective and important means that the United States can take immediately to deal with the climate crisis. The fourth climate policy of the Biden administration is to commit to preserve 30 percent of America's lands and waters by 2030, to enhance biodiversity and nature's resilience, and to contribute to the well-being of all human beings while minimizing possible economic side effects. The fifth climate policy of the Biden administration is to attach importance to science and to take it as the basis for government actions on climate change. For example, the social cost of carbon reveals how much it costs governments to deal with the consequences of emitting carbon into the atmosphere, or tells policymakers how much money they can save by reducing carbon emissions. The Obama administration introduced the first estimated social cost of carbon and it was US $50 a tonne. The Trump administration estimate was US $1 a tonne. One of the executive orders signed on Biden's first day in taking office directed federal agencies to re-estimate the social cost of carbon, requiring interim results within 30 days and final data within a year. In addition, Biden emphasized environmental justice and planned to establish the White House Environmental Justice Interagency Council (IAC) and the White House Environmental Justice Advisory Council (WHEJAC) to address current and historical environmental injustices, promising to deliver 40 percent of the overall benefits from federal government-related investments to disadvantaged communities. He also planned to establish the Interagency Working Group on Coal and

Power Plant Communities and Economic Revitalization (IWG), co-chaired by the National Climate Adviser and the Director of the National Economic Council. This organization will be responsible for guiding the just transition of fossil fuel industries toward a clean economy, reducing pollution at the source, increasing employment, and restoring economic vitality.

Climate policies of the United States have swayed several times for regime changes. Political disputes between the two parties in America have also made it more difficult and uncertain to implement climate policies. It remains to be seen whether climate policies of the Biden administration will be effectively implemented and turned into concrete actions.

53. What Contributions Have China Made to Global Climate Governance?

As the world's largest developing country, China emits about 29 percent of the total global emissions. We have taken concrete actions to tackle climate change, actively and constructively participated in global climate governance, proposed China's action plans, and contributed China's wisdom, which shows our charisma of being a responsible major country.

China been an active participator in global climate governance negotiations. We have been actively participating in the process of international governance related to climate issues. For example, we have played a constructive role in negotiations on the United Nations Framework Convention on Climate Change and sent personnel to participate in various international processes outside the Convention, such as the Forum on Millennium Development Goals, the Major Economies Forum on Energy and Climate, the International Civil Aviation Organization, the International Maritime Organization, the High-Level Advisory Group on Climate Change Financing chaired by the United Nations Secretary-General, and other cooperation mechanisms.

China has been actively advancing international climate cooperation. As a major country, we have been giving full play to our influence, strengthening communication and coordination with all parties, and promoting global climate governance. On one hand, China has been maintaining close communication with other countries to seek consensuses. For example, we have successively issued several joint statements on climate change with the United States, the United Kingdom, India, Brazil, the European Union, and France, and reached a series of consensuses on strengthening cooperation on climate change and advancing multilateralism; we have also played a constructive and leading role among developing countries and safeguarded the unity and common interests of developing countries through such negotiating groups as the BASIC countries, the Like Minded-Group of Developing Countries, and the Group of 77 and China. On the other hand, China has been helping other developing countries that are seriously affected by climate change and have weak coping capacity. Over the past decades, we have been actively supporting African countries, small island states, and least developed countries to enhance their capacity to cope with climate change through various channels of climate change cooperation, such as the China South-South Climate Cooperation Fund, the Ten-Hundred-Thousand Project (namely implementing 10 low-carbon demonstration zones, 100 climate change mitigation and adaptation projects in developing countries, and providing 1,000 people with climate change training), the Belt and Road Initiative, etc.

In response to climate change, China has set a good example to other countries by taking pragmatic climate actions at home. We have taken the initiative to take climate actions and to realize low-carbon development, and achieved remarkable results. These actions include: 1. Establishing the concept of ecological civilization, taking it as a priority, and integrating it into every aspect and the whole process of economic construction, political construction, cultural construction, and social construction; 2. Optimizing industrial structure. In China, the tertiary industry has surpassed the secondary industry to become the leading industry, and the proportion of the three industries was 7.1:38.6:54.3 in 2019; 3. Adjusting energy structure. In 2019, the share of coal consumption in total energy consumption decreased to 57.7 percent, and the share of non-fossil fuel energy increased to 15.3 percent; 4. Improving energy conservation and energy efficiency. In 2019, the energy consumption per unit of GDP was 0.49 tonnes of standard coal/10,000 yuan; 5. Vigorously advancing afforestation as well as ecological construction and protection. China has also proposed medium and long-term emissions reduction targets, such as achieving carbon peaking before 2030; lowering carbon dioxide emissions per unit of GDP by over 65 percent from the 2005 level, increasing the share of non-fossil fuels in primary energy consumption to around 25 percent, increasing forest stock by 6 billion cubic meters above the 2005 level, and bringing the total installed capacity of wind and solar power to over 1,200 GW by 2030; striving to achieve carbon neutrality before 2060.

54. What Is the Significance of International Cooperation to Global Actions on Climate Change?

Addressing climate change is a global public issue. The Earth's atmospheric resources are public goods, and impacts of climate change and climate governance are global, which determines that climate change can hardly be effectively tackled by a single country.

Though international cooperation, the international community sets out goals and pathways for actions on climate change. On one hand, international cooperation can enhance climate consciousness and scientific and technological innovation. Through exchanges and cooperation, the international community can enhance its understanding of climate issues and clarify action goals, and promote the development and popularization of climate-friendly technologies. On the other hand, through international cooperation, the international community can guide the directions of investments, markets, and economic development. By means of various modes of financial assistance and international trade rules, the world can establish a climate and environment-friendly market system to better guide the establishment of a low-carbon economy.

The United Nations Framework Convention on Climate Change and other mechanisms established for international cooperation have become effective platforms for global cooperation in climate governance. Negotiations on climate action targets under platforms of the United Nations, as well as climate dialogues under the G20, APEC, and other relevant international mechanisms can help countries around the world reach consensuses and take effective climate actions. Countries in different development stages have different capabilities to tackle climate change. International cooperation can help more countries better achieve low-carbon transition and development while ensuring global climate security.

Section III: Pathways for Transformation and Development Under the Vision of Achieving Carbon Peaking and Carbon Neutrality

55. What Targets Has China Set to Tackle Climate Change Since the 11th Five-Year Plan? How Have These Targets Been Implemented?

China has always been attaching great importance to climate change, and has taken proactive response to climate change as a major strategy for national socioeconomic development. Since the 11th Five-Year Plan, China has set targets to tackle climate change, and China's State Council has formulated and implemented comprehensive work plans for energy conservation and emissions reduction. For example,

Energy intensity targets during the 11th Five-Year Plan period (2006–2010): The concept of energy conservation and emissions reduction, and binding indicators for the target of reducing energy consumption per unit of GDP by about 20 percent from the level by the end of the 10th Five-Year Plan period were stated in the 11th Five-Year Plan for the first time. During the 11th Five-Year Plan period, China's energy consumption per unit of GDP decreased by 19.1 percent, basically completing the objectives and tasks set out in the Outline of the 11th Five-Year Plan for Economic and Social Development.

Carbon dioxide intensity targets during the 12th Five-Year Plan period (2011–2015): The targets for increasing the use of low-carbon energy and reducing fossil fuel consumption was stated in the 12th Five-Year Plan: increasing the share of non-fossil fuel energy in primary energy consumption to 11.4 percent; reducing energy consumption per unit of GDP by 16 percent; reducing carbon dioxide emissions per unit of GDP by 17 percent; increasing the percentage of forest cover to 21.66 percent and forest stock by 600 million cubic meters. During the 12th Five-Year Plan period, China achieved extraordinarily remarkable results, with a cumulative 20 percent reduction in carbon intensity, the share of non-fossil fuel energy in primary energy consumption reaching 12 percent in 2015, the installed capacity of renewable energy accounting for a quarter of the world's total, and the newly added installed capacity of renewable energy accounting for a third of the world's total. China has made proactive contributions to global actions on climate change.

Targets for controlling total energy consumption and energy intensity during the 13th Five-Year Plan period (2016–2020): Binding indicators for tackling climate change include: increasing the share of non-fossil fuel energy in primary energy consumption to 15 percent; reducing energy consumption per unit of GDP by 15 percent; reducing carbon dioxide emissions per unit of GDP by 18 percent; increasing the percentage of forest cover to 23.04 percent and the forest stock by 1.4 billion cubic meters. It was stated in The 13th Five-Year Comprehensive Work Plan for Energy Conservation and Emissions Reduction that

China's energy consumption per 10,000 yuan of GDP should decline by 15 percent from the 2015 level, and the total energy consumption should be controlled within 5 gigatonnes of standard coal. According to the Department of Energy Statistics of the National Bureau of Statistics of China, the total energy consumption in 2020 was about 4.97 gigatonnes of standard coal, completing the target of controlling the total energy consumption within 5 gigatonnes of standard coal set out in the Outline of the 13th Five-Year Plan for Economic and Social Development. However, China's cumulative energy consumption intensity declined by only about 13.79 percent, underachieving the target of reducing energy consumption per unit of GDP by 15 percent from the 2015 level set out in the Outline of the 13th Five-Year Plan for Economic and Social Development. CO_2 emissions per unit of GDP declined by about 22 percent, overachieving the 18 percent target set out in the Outline of the 13th Five-Year Plan for Economic and Social Development.

The 14th Five-Year Plan period is very crucial for achieving carbon peaking and carbon neutrality. The first year of this period falls in 2021, when China planned to formulate national and provincial action plans to achieve carbon peaking, and to clarify pathways and policies to ensure carbon peaking before 2030 and lay a solid foundation for achieving carbon neutrality before 2060. It was stated in the 14th Five-Year Plan that the energy consumption and carbon emissions per unit of GDP should decline by 13.5 percent and 18 percent respectively from the 2020 levels by 2025. The Plan for Controlling Energy Consumption Intensity and Total Consumption, released in September 2021, is expected to advance the steady shift from controlling energy consumption intensity and total energy consumption to controlling carbon consumption intensity and total carbon consumption.

On the whole, there are certain rules behind China's targets for tackling climate change. On one hand, China has shifted from controlling relative emissions (both energy intensity and carbon intensity) to controlling carbon consumption intensity and total carbon consumption by strengthening the control over energy consumption intensity and total energy consumption, so as to achieve carbon peaking and carbon neutrality by controlling absolute emissions. On the other hand, China has continuously shifted the control over fossil fuel consumption to an all-round development layout covering the development of non-fossil fuel energy, forest carbon sinks, industrial and regional climate change adaptation, etc. China's actions on climate change have been steadily advanced at national and local levels, and remarkable results have been achieved.

56. Is It Possible for China to Achieve Carbon Peaking Ahead of Schedule During the 14th Five-Year Plan Period?

China has been unswervingly implementing the national strategy of actively tackling climate change, participating in and leading global climate governance, and effectively promoting the construction of ecological civilization and environmental protection. During the 13th Five-Year Plan period, China actively implemented the national strategy of tackling climate change by adopting a series of measures, such as industrial restructuring, optimizing energy structure, improving energy conservation and energy efficiency, promoting the construction of carbon markets, and increasing forest carbon sinks. These climate actions produced remarkable results. Contrary to the previous trend that China's total carbon dioxide emissions increased at a fast speed, the constant decline in carbon dioxide emissions per unit of GDP has laid a solid foundation for China to achieve net zero emissions ahead of schedule.

By the end of 2019, China's carbon intensity had decreased by 18.2 percent from the 2015 level and 48.1 percent from the 2005 level; the share of non-fossil fuel energy in energy consumption had reached 15.3 percent, overachieving the targets that China had promised internationally to achieve by 2020. In addition, China's energy consumption per unit of industrial added value of enterprises above designated size decreased by more than 15 percent from the 2015 level, saving 480 million tonnes of standard coal equivalent; China's green buildings accounted for 60 percent of new residential buildings in cities and towns; new energy vehicles accounted for 55 percent of global new energy vehicles, making China the country with the largest stock of new energy vehicles in the world. China has gradually gained an advantage in developing and utilizing renewable energy. According to statistics of the electric power industry, by the end of October 2021, China's cumulative installed capacity of renewable energy had exceeded 1,000 GW, doubling the 2015 level; it accounted for 43.5 percent of the total installed capacity of power generation in China, 10.2 percent higher than the 2015 level; it accounted for about 32 percent of the world's total renewable energy power generation, with hydropower, wind power, photovoltaic power, and biomass power generation all ranking first in the world. With top-ranking technologies and manufacturing scale, China has formed a complete industrial chain for wind and photovoltaic power generation equipment manufacturing. We have manufactured about 41 percent of the world's total wind turbines.

In an address to the United Nations General Assembly, President Xi announced that China aimed to achieve carbon peaking before 2030 and carbon neutrality before 2060. In 2020, the Central Economic Work Conference deployed the task named "Being Well-Prepared for Achieving Carbon Peaking and Carbon Neutrality" as one of the eight tasks in 2021. The Ministry of Ecological Environment released the Guiding Opinions on Overall Planning and Strengthening the Work Related to Climate Change and Environmental Protection (hereinafter referred to as "the Opinions") to provide support for achieving carbon peaking and carbon neutrality. The Opinions stated that the Ministry of Ecological Environment would hasten to formulate action plans for achieving carbon peaking before 2030 and comprehensively adopt relevant policy tools and instruments for advancing implementation. Under the objective of achieving carbon peaking before 2030 and the vision of achieving carbon neutrality before 2060, industries in China have been strengthening efforts for energy conservation and emissions reduction. Shanghai, Hainan, Jiangsu, Guangdong, and other places have successively proposed to take the lead in achieving carbon peaking. Many enterprises, such as the State Power Investment Corporation, the China Energy Investment Corporation, and the Datang Group, have clearly stated that they would achieve carbon peaking ahead of schedule. Therefore, the whole nation has been galvanized to take actions to achieve carbon peaking.

Generally speaking, the energy transition is the basis for achieving carbon peaking. At present, the annual growth rate of fossil fuel energy consumption in China is about 2 percent, while the growth rate of hydropower, nuclear power, and wind power generation is about 10 percent. The increments of fossil fuel consumption can be completely offset by increasing low-carbon energy, including renewable energy, nuclear power, natural gas (a relatively low-carbon fossil fuel), and other energy sources. China has set the goal of achieving carbon peaking before 2030. But if we can hasten to adjust and optimize industrial structure and energy structure, reach peak coal as soon as possible, vigorously develop renewable energy, and realize the transformation of energy structure, industrial structure, consumption behaviors and consumption content ahead of schedule, it is possible to achieve carbon peaking ahead of schedule during the 14th Five-Year Plan period.

57. What Acute Challenges China Has to Face in Achieving Carbon Neutrality?

By strengthening policy guidance and control, China is possible to achieve carbon peaking ahead of schedule during the 14th Five-Year Plan period, which will lay a solid foundation for China to achieve carbon neutrality. But as the largest developing country in the world, China still faces acute challenges in achieving carbon neutrality before 2060. For the urgency and toughness of the task, we need to make relentless efforts for the following reasons.

Firstly, in terms of total emissions, China's total carbon emissions are huge, accounting for about 28 percent of the world's total, more than twice those of the United States and more than three times those of the European Union. Therefore, China needs to reduce more carbon emissions than other economies to achieve carbon peaking. Secondly, in terms of development stage, with mature economies, European countries and America have achieved absolute decoupling between economic development and carbon emissions, and have seen a steady decline in carbon emissions. China's GDP ranks second in the world, while the GDP per capita has just exceeded US $10,000. We still have to face the acute problem of unbalanced and inadequate development, which requires massive energy supply. Therefore, China has not yet achieved carbon peaking. It is very difficult for China to coordinate socioeconomic development, structural transformation, the low-carbon energy transition, and the goal of achieving carbon peaking and carbon neutrality. Thirdly, in terms of the trend in carbon emissions, developed European countries, such as Britain, France, and Germany, have achieved carbon peaking as early as 1990 long before international climate negotiations; the United States, Canada, Spain, Italy, and other countries have achieved carbon peaking around 2007. Then it will take them 40 years, 60–70 years, or even longer years to achieve carbon neutrality by 2050. However, it will only take China 30 years to achieve carbon neutrality before 2060 after achieving carbon peaking before 2030, significantly faster than European countries and the United States. So, China has to pay more efforts and hasten to achieve carbon neutrality. Fourthly, in terms of key industries and fields, coal is the dominant energy source in China's energy structure. In 2019, China's coal consumption accounted for 57.7 percent of the total energy consumption, non-fossil fuel energy consumption accounted for 15.3 percent of the total energy consumption, and thermal power generation accounted for 72 percent of the power generation of all power plants above designated size. China can hardly phase out 85 percent of fossil fuels in only

30 years to achieve zero carbon emissions through energy conservation and emissions reduction alone. We need a truly clean energy revolution. At present, a new round of energy revolution characterized by clean and low-carbon production is booming in China, and replacing traditional energy with new clean energy is an overwhelming trend. China must revolutionize the energy system before achieving carbon peaking and carbon neutrality. The goal of achieving carbon peaking before 2030 and achieving carbon neutrality before 2060 requires a radical transformation of the energy system in the near future so that China can achieve negative emissions after achieving net zero emissions around 2050.

In the future, China's total energy consumption will maintain a steady growth, with a consumption of about 6.5 gigatonnes of standard coal by 2050. In terms of energy structure, by 2050, the share of non-fossil fuel energy in total energy consumption will reach nearly 80 percent, and the share of coal, oil, and natural gas in total energy consumption will decrease to less than 9 percent, 2 percent, and 9 percent respectively. By applying carbon capture and storage to fossil fuel-fired power generation and heat supply as well as large boilers, and adopting negative emissions technologies, such as bioenergy with carbon capture and storage (BECCS) and direct air carbon capture and storage (DACCS), China is expected to achieve net zero CO_2 emissions from energy activities and negative emissions of electricity systems by 2050.

If so, end-use sectors must advance high levels of electrification to reduce carbon emissions in the context of achieving carbon neutrality. Given a fast decline in the CO_2 emissions coefficient of electricity, more electricity consumption means more efforts to reduce emissions until negative emissions by 2050. Among end-use sectors, the industrial sector, the transportation sector, and the building sector should take more efforts to advance electrification.

To realize the transformation of the energy system, China needs to take immediate actions to vigorously promote the rapid development of non-fossil fuel energy, such as renewable energy and nuclear power, so as to control the use of fossil fuels. During the 14th Five-Year Plan period, China needs to endeavor to reach peak coal and oil. We also need to reach peak gas around 2035. To realize the transformation of the energy system, China also needs to develop carbon capture and storage. The goal of achieving carbon peaking and carbon neutrality requires China to foster scientific and technological innovation and adjust overall technological layout and development direction.

58. In the Context of Achieving Carbon Peaking and Carbon Neutrality, What Acute Challenges China Has to Face amid Coal Phase-out?

China is the world's largest producer and consumer of coal. In 2020, China accounted for about 50 percent of the world's coal production and up to 54.3 percent of the world's coal consumption. According to the China National Coal Association, 6 coal enterprises had an annual output of more than 100 million tonnes, and 4 coal enterprises had a production capacity of more than 200 million tonnes. Compared with oil and natural gas, coal is the fossil fuel with the highest carbon intensity. The amount of CO_2 emitted per unit of heat content from coal is about 1.31 times that of oil and 1.72 times that of natural gas. To achieve carbon peaking and carbon neutrality, China still faces acute challenges from the green and low-carbon transition of the coal industry chain, which encompasses coal development, utilization, and transformation.

Upon the 2015 Paris Climate Conference, the World Wide Fund for Nature (WWF) predicted that if we were serious about the Paris Agreement and equipped all coal-fired plants with the most efficient technologies, they would inevitably produce more emissions than the 1.5 degrees Celsius target. Coal has no future! Large-scale application of carbon capture, utilization and storage (CCUS) can theoretically reduce carbon emissions from coal consumption. However, applying this end-of-pipe technology is subject to technological factors, economic factors, environmental factors, the matching between emission sources and storage conditions, and other factors. With slow progress in deploying CCUS, we must vigorously control and reduce coal use and phase out coal power.

In 2017, the United Kingdom, Canada, and other countries jointly launched the Powering Past Coal Alliance to actively advance the process of phasing out coal globally. At present, more than a dozen of European countries which have basically completed coal phase-out have actively participated in the alliance and formulated a timetable for complete coal phase-out. Britain is the first country in the world to use coal power. In 2017, the share of coal power in Britain was only 7 percent. During her visit to Canada, the then British Prime Minister, Theresa May, announced that Britain would completely phase out existing coal power before 2025, requiring that the instantaneous carbon intensity of any power plant should not exceed 450 g/kWh from October 1, 2025 onward. Germany is a large coal consumer in Europe, with coal power accounting for 40 percent of the total power generation in 2016. So far, the

Germany federal government and relevant energy companies have formally signed a lot of agreements to stop using lignite as a raw material for power generation by 2038.

In recent years, China has achieved remarkable results in controlling the total amount of coal consumption. In 2019, coal accounted for 57.7 percent of China's total energy consumption, down by 10.8 percent from the 2012 level. Since 2016, more than 800 million tonnes of cumulative overcapacity have been phased out, and 65 GW of coal power capacity were eliminated, halted, and shelved in 2017 alone. On one hand, as the stabilizer and ballast for China's energy security, coal power plays an important role in supporting China's economic development. On the other hand, the goal of achieving carbon peaking and carbon neutrality is a great challenge for the coal industry. Once being a dominant source of electricity generation, coal power has now played an important role in peak shaving, frequency control, and voltage regulation, and serving as a backup source of electricity generation, thus complementing renewable energy perfectly. In the transformation of the coal supply chain, thoughtful arrangement is needed for millions of redundant labor force in the coal industry. With recessions and declines, some resource-dependent cities and regions may face great difficulties in transformation. A large number of coal-related infrastructure, including advanced thermal power units, have to be phased out ahead of schedule. An increasing number of financial institutions have also announced that they will no longer invest in coal power projects in the context of achieving carbon peaking and carbon neutrality, posing great risks to a large number of China's overseas coal power projects.

Producing specific chemicals from coal is also an important way of coal conversion and utilization. Both traditional coal-derived chemicals and modern coal-derived chemicals will emit carbon dioxide in the process of coal conversion. Although the target of achieving carbon peaking does not involve carbon emissions in industrial processes, the goal of achieving carbon neutrality does involve emissions from coal conversion. Coal-derived chemicals are greatly subject to international oil prices. In the carbon-constrained world, we need to make comprehensive assessments on impacts of coal-derived chemicals on water, energy consumption, and the environment, so as to find proper pathways for future transformation and development.

In a word, the coal industry must accelerate transformation amid the great challenge of achieving carbon peaking and carbon neutrality. It must delve into specific pathways to achieve a steady transformation.

59. How Does Natural Gas Contribute to Carbon Peaking and Carbon Neutrality?

Natural gas emits less carbon dioxide than coal and oil, but it is also a carbon energy source. In 2019, the International Energy Agency (IEA) released *The Oil and Gas Industry in Energy Transitions*. This report stated that energy intensity reduction, alternative low-carbon fuels, and the rapid development of renewable energy have slowed down the growth rate of global carbon emissions. Since 2010, the world has reduced 500 million tonnes of carbon dioxide by replacing coal with natural gas, which plays a key role in promoting energy transitions and is a useful transitional energy source in the process of energy transitions. In Europe, the process of replacing coal with natural gas has been basically completed, and the consumption of natural gas has reached a plateau. Some European countries, such as France and Netherlands, have stopped using natural gas as a clean energy source and tried to reduce the use of natural gas. In January 2019, the Institute for Sustainable Development and International Relations (IDDRI), a think tank in France, released the report *Natural Gas and Climate Commitments, Two Irreconcilable Factors?*. According to the research report, 6 $GtCO_2$ emitted from natural gas combustion accounted for about 12 percent of global greenhouse gas emissions in 2017. To achieve carbon neutrality, the European Union must establish a common vision of reducing the use of natural gas substantially and take corresponding policy actions. On January 20, 2021, Werner Hoyer, the President of the European Investment Bank, said at the annual meeting that Europe needed to ensure that fossil fuels would not be used any longer, that is to say, the era of natural gas was over.

In 2019, China's natural gas output (including unconventional gas) was 177.3 billion cubic meters, natural gas imports 135.2 billion cubic meters, and apparent consumption 306.4 billion cubic meters, which accounted for 8.1 percent of primary energy consumption. In December 2020, the State Council Information Office of China issued a white paper titled "Energy in China's New Era". It stated a series of requirements for the future development of China's natural gas industry, including replacing coal with natural gas, strengthening the development and connectivity of natural gas infrastructure, and making the use of natural gas more efficient in urban areas, industrial fuels, power generation, transport, and other fields. China should also vigorously promote the use of gas-fired combined cooling, heating, and power (CCHP) system, the development of decentralized renewable energy, and coordination of multiple energy sources and energy cascade use in final energy consumption.

China should be proactive to integrate biogas into industrial development and encourage methane upgrading in rural areas. China should also reasonably deploy and properly develop natural gas power generation, make use of its flexibility, and have it play an important role in peak shaving. The National Energy Administration proposed six key areas that should be highly noticed and stated that China should further innovate development patterns, accelerate the development and utilization of clean energy, make non-fossil fuel energy and natural gas major sources of energy consumption increments, and greatly increase the share of clean energy consumption during the 14th Five-Year Plan period. Therefore, in the context of achieving carbon peaking and carbon neutrality, natural gas has huge potential as a transitional energy source in the process of energy transitions in the next 5–10 years. However, its development prospect in the next 10–15 years remains uncertain. In the long run, natural gas is bound to be replaced by carbon-free non-fossil renewable energy sources in the context of carbon neutrality.

60. How Does Renewable Energy Contribute to Carbon Peaking and Carbon Neutrality?

Replacing fossil fuels with renewable energy will contribute a lot to the transformation of energy system. Carbon peaking and carbon neutrality means that in a long period, China must advance clean electricity generation by encouraging the development of wind power, photovoltaic power, and other renewable energy sources.

During the 13th Five-Year Plan period, China's installed capacity of renewable energy grew by an average of 60 GW annually, at an incomparable rate of 32 percent. By the end of 2019, China's installed capacity of renewable energy had reached 414 GW, accounting for 20.63 percent of the total installed power capacity; China's wind power generation capacity was 210 GW, and photovoltaic power generation capacity was 204 GW; renewable energy generated 630 GWh of electricity, accounting for 8.6 percent of the total power generation. In 2020, when the world was seriously affected by the COVID-19 pandemic, the global renewable energy power generation increased by nearly 7 percent from the 2019 level, accounting for 28 percent of the global total power generation. According to the International Energy Administration, in 2020, China's newly added grid-connected installed capacity of wind power was 71.67 GW, and the newly added installed capacity of photovoltaic power was 48.2 GW, with a year-on-year growth rate of 60 percent.

At the Climate Ambition Summit 2020, President Xi said that to develop renewable energy, China should increase the total installed capacity of wind and solar power to more than 1,200 GW by 2030, which has galvanized industrial enterprises to develop renewable energy more enthusiastically. In December 2020, the National Energy Administration proposed the goal of increasing China's installed capacity of wind and solar power by 120 GW in 2021. In October 2020, representatives from more than 400 wind energy enterprises around the world jointly signed and released Beijing Declaration on Wind Energy: Developing 3,000 GW of Wind Power, Leading Green Development, and Achieving China's 2060 Carbon Neutrality Target. They solemnly proposed that "the 14th Five-Year Plan published in 2021 should outline a sufficiently ambitious target for wind power, aiming for more than 50 GW of annual installations from 2021 to 2025 and more than 60 GW annually from 2026 onwards. An industrial development plan of this scale would bring China's cumulative wind capacity to at least 800 GW by 2030 and 3,000 GW by 2060, in line with the long-term carbon neutrality goal."

To achieve carbon peaking and carbon neutrality, traditional electric power enterprises should actively develop renewable energy and strive to be the vanguard of carbon neutrality. By 2020, China had signed a series of agreements to launch wind, solar, and energy storage projects, worth of hundreds of billions of yuan accumulatively. Petrochemical enterprises had also attached great importance to renewable energy, taking it as a new field for technical and economic competition. For example, by the end of 2020, the total installed power capacity of the State Power Investment Corporation had reached 176 GW, with clean energy accounting for 56.09 percent and reaching 98.88 GW; the installed capacity of wind and photovoltaic power had reached 60.49 GW, ranking first in the world. In 2020, China Huaneng Group's newly added installed capacity of renewable energy reached more than 10 GW, exceeding the total increments of the previous four years.

The rapid development of renewable energy is conducive to the energy transition. However, it may pose great risks to the stability of power grid and environmental protection. As the "stabilizer" of power grid, pumped storage power stations, characterized by open-loop or closed-loop, quick response, bidirectional design of power conversion, and regulation of storage capacity, is a perfect technical choice to enhance the stable operation of power grids. So far, the scale of pumped storage power in operation and under construction of the State Grid Corporation of China has reached 62.29 GW, ranking first in the world. The issue of environmental protection, which is subject to widespread concern, has become one of the factors constraining the development of the renewable energy industry. In February 2021, the State Forestry and Grassland Administration released the Notice on Standardizing the Use of Forestlands for Construction of Wind Farm Projects, which clearly stated that the use of forestlands in key forestry districts for construction of wind power projects is strictly prohibited. This will make it even more difficult to develop centralized wind power projects. For example, Zhejiang, Hunan, and other provinces with good vegetation cover suspended approvals for onshore wind projects before. With the rapid development of the renewable energy industry, large-scale engineering construction may have a far-reaching impact on ecosystems, biodiversity, and communities. Governments must make a trade-off and meet growing demand for clean and cheap energy while minimize its damage to and impacts on land and water.

61. How Does Nuclear Power Contribute to Carbon Peaking and Carbon Neutrality?

After more than 30 years of development, China has become one of the world's largest producers of nuclear power. By the end of 2019, the installed capacity of nuclear power in operation and under construction had been 65.93 GW, ranking second in the world, and the installed capacity of nuclear power under construction had ranked first in the world. China has now developed a complete nuclear power industry chain, which encompasses upstream and downstream links, such as the manufacturing of nuclear power plant equipment, design and construction of nuclear power plants, operation of nuclear power plants, supply of nuclear fuels, and disposal of radioactive waste. Several nuclear power plants applying advanced third-generation technologies have been built in China, and significant breakthroughs have been made in nuclear technologies, such as the next generation of nuclear power plants and small modular reactors. The white paper Energy in China's New Era released in December 2020 stated that building a clean and diversified energy supply system, prioritizing the development of non-fossil fuel energy, and developing nuclear power in a safe and orderly way are integral parts of building a clean and diversified energy supply system. Regarding nuclear safety as the lifeline of nuclear power development, China has been adhering to the principle of paying equal attention to development and safety, implementing the strategy of developing nuclear power in a safe and orderly way, strengthening the management and supervision of the whole life cycle of nuclear power plants, including planning, site selection, design, construction, operation, and decommissioning, and sticking to adoption of the most advanced technologies and the strictest standards to develop nuclear power. So far, China's nuclear power units have been in normal operation overall, without any occurrence of incidents or accidents at the Level 2 or above classified under the International Nuclear and Radiological Event Scale. Adhering to the strategy of developing nuclear power plants in a safe way is of great strategic significance for achieving China's ambitions to build a safe and efficient energy system, to cope with challenges caused by global climate change, to ensure sustainable development, to accelerate scientific and technological innovation, and to safeguard and enhance national security.

62. How Does Energy Storage Technologies Contribute to Carbon Peaking and Carbon Neutrality?

For purpose of achieving carbon peaking and carbon neutrality, a low-carbon energy system is the key and priority. Renewable energy plus energy storage is one of the schemes to make energy systems low-carbon and flexible. Breakthroughs in energy storage technologies will contribute to dramatic declines in costs and increase the competitiveness of the economy. Their large-scale application will make the traditional one-way and linear power generation, transmission, distribution, and supply become multi-directional and annular, speed up two-way conversions of energy, electric power, and substances, and integrate electrification into every aspect of economic society.

The energy storage industry is booming around the world, and new technologies and business models continue to emerge in recent years, presenting bright future prospects. However, because of its very nature, energy storage faces systemic problems in such aspects as application, conditions, safety, technologies, business models and others. This makes it clear that energy storage cannot develop entirely on its own, independent of the development of renewable energy, demands of the electricity systems, and economic and social needs. If we continue to improve technologies and policies during the 14th Five-Year Plan period, we can advance the development of the energy industry and low-carbon transition.

Over the past decade, the cost of energy storage has fallen by an average of 10–15 percent per year, which is attributed to advances in energy storage technologies. The cost of energy storage systems has declined from 7–8 yuan/Wh to 2 yuan/Wh, and then to nearly 1.5 yuan/Wh; the cycle life of batteries has also been prolonged from 1,500 to 3,400, and even now 6,500. The decline in the costs of the whole system has in turn caused lower installed costs and cost per kilowatt-hour. At present, the cost per kilowatt-hour of a lithium-ion battery pack is about 0.53 yuan/kWh. Of course, the final price is subject to many boundary conditions, such as depth of charge and discharge, life cycle, etc. Most experts believe that when the cost drops to about 0.35 yuan/kWh, it will be economical, and renewable energy storage will be more feasible.

Energy storage is just one of the options that provides flexibility in energy systems. So, a proper assessment means that we should first conduct comprehensive and systematic studies of demand flexibility and compare energy storage with other options, such as responding to demands, retrofitting power plants, enhancing smart grids, and other technologies which can improve the overall flexibility.

Technologies are of crucial importance to low-carbon energy systems. Some experts even believe that the development of the energy storage industry has a direct impact on the breadth, depth, progress, and even success of the low-carbon transition of energy and electricity systems.

63. What Targets Does the Industrial Sector Need to Meet? How to Achieve Transformation and Development in the Context of Achieving Carbon Peaking and Carbon Neutrality?

Achieving carbon peaking and carbon neutrality requires the industrial sector to achieve carbon peaking and deep emissions reduction as early as possible. To that end, the industrial sector needs to continue to promote industrial energy conservation and greatly enhance the level of electrification. In industries which produce huge emissions, fossil fuels are still used as raw materials and process materials of many industrial products. There are two options to promote deep emissions reduction in these industries: 1. To adopt CCS; 2. To revolutionize process technologies and adopt GHG emissions-free technologies. At present, hydrogen-based industries, which use green hydrogen as raw materials or process materials in lieu of fossil fuels, are highly concerned.

In energy-intensive industries, energy conservation manifests itself in process technologies, motors, and innovative processes; electrification manifests itself in the use of electric boilers to provide heat for existing production processes. For better smog control, industrial boilers with a capacity of less than 35 t/h need to be replaced by electric boilers. Measures in response to smog control also include establishing clean energy centers in industrial parks and replacing small boilers with large boilers. In the long run, the electrification of the industrial sector is costly. However, electric heating is indeed a good option for producing some process materials.

For industries which produce huge emissions, such as steel manufacturing, cement manufacturing, and the chemical industry, near-term efforts to promote technological innovation of processes are a precursor to the long-term technological transformation of hydrogen-based industries.

At present, energy-intensive industries accounts for 70 percent of industrial energy consumption and 92 percent of industrial coal consumption. During the 14th Five-Year Plan period, China needs to promote industrial energy conservation and energy substitution to reduce coal consumption and enhance the level of electrification, so as to reach carbon emissions peak in the industrial sector before 2025. By 2050, electricity and hydrogen are expected to account for 58 percent of industrial final energy, and fossil fuels are expected to account for 36 percent, of which 10 percent will be used as raw materials. With the electrification of the industrial sector and the use of CCS and hydrogen-based production processes, the industrial sector is expected to achieve near zero emissions by 2050.

64. What Targets Does the Transportation Sector Need to Meet in the Context of Achieving Carbon Peaking and Carbon Neutrality?

In the context of achieving carbon neutrality, the transportation sector needs to achieve near zero emissions by 2050. At that time, cars and buses will be basically powered by batteries, most small and medium-sized trucks will be powered by batteries and oil, and some heavy-duty trucks will be powered by hydrogen fuel cells; small ships will be powered by batteries, and large ships will be powered by hydrogen fuel cells; railways that are hard to electrify will be powered by hydrogen fuel; small regional aircrafts will be powered by batteries, and large aircrafts will be powered hydrogen fuel. Given the amount of time it takes for hydrogen-powered aircrafts to go from research and development to commercial application, biofuels are expected to replace aviation kerosene for the existing fleet of fuel-powered aircrafts by 2050. During the 14th Five-Year Plan period, governments should continue to encourage the development of electric vehicles so that electric vehicles will be cheaper than fuel-powered vehicles and need no subsidies from 2025 onwards; some low-carbon cities should be encouraged to stop selling fuel-powered vehicles and take measures to embrace electric vehicles, such as planning dedicated lanes for buses and electric vehicles and relocating gas stations out of urban areas. Further research and development of new technologies will also be in order, such as powering vehicles with fuel cells as well as hydrogen-powered aircrafts.

In the near term, governments should vigorously encourage the development of advanced energy-efficient vehicles, electric vehicles, hydrogen-powered heavy-duty trucks, and public transport that accommodates bicycles and other modes of transport for non-motoerized mobility, so as to reach carbon emissions peak in the transportation sector around 2025.

To meet the target that the transportation sector should achieve overall electrification and be powered by hydrogen fuel cells by 2050, the aviation industry needs to partly use biofuels. In terms of road traffic, only some heavy-duty trucks need to be powered by hydrogen fuel cells, and other types of vehicles will be powered by electricity. In terms of water transportation, large ships need to be powered by hydrogen fuel cells, and other types of ships will be powered by batteries. Railways that are hard to electrify need to be powered by hydrogen fuel cells. Large aircrafts need to be powered by biofuels and hydrogen fuel, and small aircrafts will be powered by batteries.

Energy efficiency of electric vehicles needs to be enhanced. Energy consumption of electric cars needs to be reduced from 15 kWh/100 km to less than 8 kWh/100 km by 2030. Non-motorized travel needs to account for more than 35 percent of urban travel.

65. What Targets Does the Building Sector Need to Meet in the Context of Achieving Carbon Peaking and Carbon Neutrality?

In the context of achieving carbon peaking and carbon neutrality, the building sector needs to enhance energy efficiency and the level of electrification. To that end, energy-efficient buildings should be greatly encouraged before 2025, especially in rural areas; ultra-low-energy building should be universally applied in developed cities; incentive policies should be adopted to enhance energy efficiency of household appliances.

Achieving carbon neutrality requires overall electrification of buildings, including cooking. Urban heating causes huge emissions. So, CCS and other new technologies, such as renewable heating and low-temperature heating reactors, should be applied in natural gas central heating systems for achieving near zero emissions of heating. But the efficiency of low-temperature heating reactors remains to be seen in the future.

Ultra-low-energy buildings should be universally applied in the building sector and are expected to become the standard for new buildings around 2025. Energy efficiency of household appliances should be enhanced. For example, the energy efficiency ratio of air conditioners should be enhanced to about 8 by 2040.

66. What Is the Significance of Demand-Side Management to Achieving Carbon Peaking and Carbon Neutrality?

Demand-side management is also crucial for achieving carbon neutrality. Previously, we were more concerned about supply-side reforms. But now we know that demand-side reforms can also create many opportunities in terms of carbon emissions reduction. They can be implemented to stimulate consumer spending in clothing, food, housing, transportation, and other consumer fields. Enterprises can explore business opportunities in these fields with consumers by sorting out major fields to achieve carbon peaking and carbon neutrality. For example, 40 percent of China's carbon emissions come from real estate and the building sector. The steel consumption of real estate accounts for 1/4 to 1/3 of China's total steel consumption, and the cement consumption of real estate accounts for more than 1/3 of China's total cement consumption. China's annual construction waste is 1.5–2.4 billion tonnes, but the recycling rate is less than 5 percent. The Engel coefficient of urban households in China fell below 30 percent in 2015, and likewise that of rural households in 2019. After becoming a moderately prosperous society, China should pursue green and low-carbon consumption rather than extravagance, waste, and excessive consumption. Therefore, demand-side management in clothing, food, housing, transportation, and other consumer fields can create many opportunities in terms of carbon emissions reduction.

67. Without Any Control on Emissions Reduction, Can We Achieve Carbon Neutrality Through Afforestation and Carbon Capture, Utilization and Storage?

According to *The People's Republic of China First Biennial Update Report on Climate Change*, in 2014, China's total greenhouse gas emissions were 11.186 gigatonnes of carbon dioxide equivalent; forestlands absorbed 840 million tonnes of carbon dioxide equivalent, accounting for about 7.5 percent of China's annual greenhouse gas emissions; China's pilot CCS/CCUS projects in operation reduced hundreds of thousands of tonnes of carbon dioxide emissions per year. According to China's NDCs, China's peak carbon dioxide emissions must to be controlled within 10–12 gigatonnes. To that end, the total volume of peak carbon emissions from electric power, industry, construction, transportation, and other energy-related sectors must be controlled within 9–11 gigatonnes. Without any control on emissions from these sectors, achieving carbon neutrality before 2060 means that we should reduce an equivalent volume of carbon emissions as mentioned above by having more carbon sinks through afforestation and installing CCUS facilities in energy-related sectors, as well as non-carbon dioxide emissions from agriculture, forestry and land use, about 1 gigatonnes of carbon dioxide equivalent. According to the studies of the increase in the amount of carbon sequestered by China's terrestrial ecosystems in the past decades, a 60 percent increment of the carbon stock of artificial forests is attributed to the increase in forest area. Given a limited increase in the area of China's artificial forests in the future, it can be predicted that China's terrestrial ecosystems can only absorb a limited amount of carbon, even if they will be effectively managed to sequester more carbon. Before 2020, China's CCUS was basically in the research and development stage or the trial stage, and reducing emissions was costly. But it does contribute a lot to carbon neutrality, which will be amplified from 2030 onwards. According to positive scenarios, CCUS is expected to contribute to about 30 percent of carbon emissions reduction by 2050. So, carbon neutrality can never be achieved through afforestation and CCUS alone.

68. What Impacts Does Carbon Neutrality Have on Employment?

To achieve carbon peaking, we must continue to strengthen efforts to reduce carbon emissions, which requires corresponding adjustments to the entire economic structure, industrial structure, and energy structure. These adjustments will affect every aspect of economic life, including consumption patterns, energy production and utilization, the development of energy technologies, industrial layout, commodity production and distribution, to name but a few. Relevant policies and targets will undoubtedly have a significant impact on employment. In the context of achieving carbon peaking, some traditional fossil fuel industries will face great pressure, and likewise some downstream industrial sectors with fossil fuels as the main raw materials. But other industries, such as the industry specializing in renewable energy utilization and energy conservation services, will embrace new development opportunities.

Policies on carbon emissions reduction and low-carbon technologies are important for peaking carbon emissions. With an ultimate aim of achieving sustainable development and tackling climate change, low-carbon development refers to the development pathway under which the transition towards a low-carbon economy is completed through low carbon emissions. Different policy objectives and pathways for policy implementation will have different effects on relevant industries. Adjusting industrial structure and energy structure, eliminating backward production capacity, developing new technologies and renewable energy can affect industrial and regional employment structure. Apart from these effects on total employment and employment structure, employees also have to sharpen their skills.

Policies on carbon emissions reduction involve many industrial sectors, and they will have significantly different impacts on employment in different sectors and regions. When setting targets for emissions reduction, governments should take into account technical feasibility, economic costs, and impacts of policies on employment. Low-carbon policies will create new investment opportunities for some sectors and increase employment. For example, investments in the development and utilization of renewable energy, increasing carbon sinks through afforestation, and professional energy conservation services will create a large number of direct and indirect employment opportunities for relevant industries, such as relevant equipment manufacturing, forest tourism, and green finance. On the contrary, coal, oil, and gas extraction and related chemical industries, thermal power, steel, cement, and other industries with high consumption of coal will have fewer employment opportunities. Low-carbon policies will have different impacts on different sectors and employment of each sector in terms of both duration and severity.

By having a structural impact on energy demand and affecting energy prices and costs, various policies on carbon emissions reduction will directly cause an increase or decrease in employment opportunities in energy-related sectors and enterprises in the short term. Specifically, some carbon-intensive sectors will gradually have fewer employment opportunities or slower employment growth, while some low-carbon industries will have more employment opportunities. Therefore, in the short term, climate policies will first significantly affect employment in those directly related industries, and specific net effects will be dependent on the total employment scale and labor productivity of these industries. In the medium term, the impacts that policies on carbon emissions reduction will have on employment will spread from carbon-intensive sectors and low-carbon sectors to the whole economy due to close connections of different sectors. For example, job losses in the coal mining and washing industry will affect the transportation sector and lead to a decrease in employment opportunities. On the contrary, the development and utilization of renewable energy, such as wind and solar energy, will create more jobs in related equipment manufacturing sectors. In the long term, a large amount of money will be invested in low-carbon technologies in the context of achieving carbon peaking, and technological progress/innovation will then create more jobs. Policies on carbon emissions reduction will promote the upgrading of employment structure in the long term, thus attracting labor force from carbon-intensive sectors to work in low-carbon sectors. Studies of the impacts of low-carbon policies and climate policies on employment have also proved that one of the long-term net effects of these policies manifests itself in more green, environmentally-friendly, and decent jobs.

Seen from impacts of these policies on specific industries, the development of clean energy and renewable energy will create a large number of new employment opportunities for the energy industry and related equipment manufacturing sectors. Vigorous development of hydropower, wind power, solar energy, bioenergy, and other renewable energy sources will create more employment opportunities for such industries as the development of renewable energy technologies and equipment manufacturing, installation, and maintenance. Energy conservation and emissions reduction need financing and financial services, thus stimulating employment in the financial industry and creating new jobs in the fields related to financial services, such as clean energy investment, trade in low-carbon technology services, clean development mechanisms, and carbon trading markets. To enhance the utilization efficiency and clean use of fossil fuels, electric power, transportation, construction, metallurgy, the chemical industry, the petrochemical industry, the automotive industry, and other industries will hasten the development of low-carbon and energy-efficient technologies. Research and development as well as application of low-carbon technologies will greatly stimulate employment at technical and service levels. Energy consulting companies and energy service companies will contribute to forming a new employment structure. Afforestation and ecological protection will also create a large number of job opportunities. Moves such as transforming farmlands to forests, setting up nature reserves, and developing eco-tourism, will increase the employment demand in forestry, horticulture, forest management, and maintenance of facilities in tourist areas. Huge investments in low-carbon development will also create a large number of employment opportunities for the service industry, including consulting, insurance, commercial meteorological services, environmental protection and popular science, and media. New jobs, such as carbon emissions management, will also emerge. In January 2021, the Ministry of Human Resources and Social Security of the People's Republic of China released the Announcement on Publicizing the Occupational Information on Technical Personnel in the Field of Integrated Circuit Engineering, which stated the plan to add 18 new occupations and adjust 21 existing occupations. It is noteworthy that carbon emissions management is one of the newly added occupations.

In the context of low-carbon transition, the development of new technologies and new industries requires the labor force to enhance their skills, which is disadvantageous to low-wage workers. As a developing country, China needs a large amount of money and a large number of high-tech talents to develop renewable energy and improve energy efficiency. With little or no knowledge about technologies, most workers in developing countries are low-paid and have fewer employment opportunities. Energy conservation and emissions reduction will limit the development of enterprises with high consumption of resources and energy, high pollution, and high emissions. Many small and medium-sized enterprises will be gradually shut down for violations of relevant regulations, thus causing some industries to shrink and resulting in structural unemployment. For example, low-paid technicians in the chemical industry, machinery manufacturing, and the steel industry may lose their jobs amid technological upgrading, small enterprises with high energy consumption and high pollution in the coal industry will be shut down, and some low-paid technicians in the construction industry, the traditional automotive industry, and the energy industry will also lose their jobs.

69. Do Different Regions Have to Achieve Carbon Peaking and Carbon Neutrality Simultaneously?

As a large developing country, China has different natural resources endowment and socioeconomic development in different regions. During the 12th Five-Year Plan period, different regions were allocated with different targets for energy conservation and emissions reduction. 31 provinces of China were divided into five groups. Provinces in eastern China were allocated with more challenging targets, and provinces in central China and western China were allocated with less challenging targets. It is impossible for different regions to achieve carbon peaking and carbon neutrality simultaneously. So, they can have different timetables for achieving green and low-carbon development in the context of achieving carbon peaking and carbon neutrality. They can adopt tailor-made policies and measures to avoid a one-size-fits-all approach. To implement important policies, China often adopts pilot schemes in specific regions and then apply them to more regions after learning by trial and error. This practice can also be applied to hasten actions on achieving carbon peaking and carbon neutrality. The Central Economic Work Conference has once mentioned encouraging and supporting regions with favorable conditions to take the lead in achieving carbon peaking. Eastern China is more developed and thus has more favorable conditions of taking the lead in achieving carbon peaking; western China abounds in renewable energy resources, such as hydropower, wind energy, solar energy, and geothermal energy. Achieving carbon peaking and carbon neutrality is a long-term task. All regions should actively explore appropriate development pathways and countermeasures for achieving carbon peaking and carbon neutrality based on their own conditions and advantages.

But when they take advantage of their own strengths to actively reduce emissions, they should not try to outdo others just for the sake of outdoing others, and they certainly must refrain from hurting the interests of other regions. At the national level, China should improve mechanisms and support policies for coordinated regional development. In some places, power cuts are imposed during year-end evaluation of government performance, in the name of energy conservation and emissions reduction, but with adverse impact on enterprises and residents. Even if such practice can reduce emissions in the short term, it goes against the animating principle behind green and low-carbon development, and would only damage the reputation and credibility of governments in relation to their efforts to achieve carbon peaking and carbon neutrality.

70. What Special Responsibilities Do Cities Have in the Context of Achieving Carbon Peaking and Carbon Neutrality?

The world is undergoing urbanization. According to *World Cities Report 2020: The Value of Sustainable Urbanization* released by UN-Habitat in December 2020, the global urbanization rate was 56.2 percent and was expected to reach 60.4 percent by 2030. Population growth will slow in highly urbanized regions. Less developed regions of East Asia, South Asia, and Africa, particularly India, China, and Nigeria, often have the fastest urbanization rate.

As the hub for human activities and economic activities, cities consume a large quantity of fossil fuels and are the main source of carbon dioxide emissions. The growth of urban carbon dioxide emissions is closely correlated with urban economic growth, human migrations, population density, industrialization levels, capital investments, and other factors. At present, the global urban population accounts for less than 60 percent of the total population, and the global urban energy consumption and greenhouse gas emissions account for about 75–80 percent of the world's total. Consumption-based emissions from nearly 100 big cities around the world accounted for 10 percent of global greenhouse gas emissions, according to a study released in June 2019 by C40, an international coalition of cities dedicated to tackling climate change. Without taking urgent actions, this figure is expected to nearly double by 2050. Therefore, cities should assume their due responsibilities for achieving carbon peaking and carbon neutrality. They have unique advantages in technologies, economy, environmental awareness, and social mobilization. As of September 2019, more than 100 cities around the world have committed to achieve net zero carbon emissions by 2050; some cities, such as Melbourne, Copenhagen, and Stockholm, have continued to take more aggressive policy actions and proposed more ambitious targets.

71. What Achievements Have Been Made in China's Low-Carbon Cities Which Have Applied Pilot Schemes?

Since 2010, the National Development and Reform Commission has applied pilot schemes in China's low-carbon provinces, autonomous regions, and cities in three batches, totaling 87 provinces, cities, autonomous regions, and counties. Years of explorations and development have witnessed remarkable achievements.

Firstly, areas which have applied pilot schemes have played a leading role in achieving low-carbon development and contributed to the transformation of development modes. Years of explorations and efforts have witnessed that carbon dioxide emissions per unit of GDP in cities which have applied pilot schemes generally decline faster compared with those which haven't, and carbon intensity in cities which have applied pilot schemes also declines significantly faster than the national average, indicating that cities which have applied pilot schemes have achieved remarkable achievements in industrial transformation, energy transformation, and enhancement of development quality and efficiency. Because cities which have applied pilot schemes have put forward stricter targets for carbon intensity reduction and taken the lead in proposing targets for carbon peaking and road maps, thus contributing to the transformation of industrial structure, optimization of energy structure, technological progress and innovation, and lifestyle changes.

Secondly, areas which have applied pilot schemes have greatly improved public understanding and capacity building of low-carbon development. By having a more scientific understanding of the concept of low-carbon development, these areas have known how to promote green and low-carbon socio-economic development simultaneously, thus contributing to the shift of the traditional concept of extensive development. They have also enhanced their capabilities of analyzing basic data and studying pathways in terms of economy, society, energy, carbon emissions, and environmental protection. Governments, enterprises and the public in these areas have enhanced their awareness of green and low-carbon development, thus laying a solid foundation for promoting green and low-carbon development.

Thirdly, a number of good practices and experiences have ensued. Cities which have applied pilot schemes have made great efforts to achieve green and low-carbon development and made remarkable achievements in industrial transformation, energy transformation, technological progress, advocacy of low-carbon lifestyles, and innovation of institutions and mechanisms, promoting green and low-carbon development, and strengthening the construction of ecological civilization. For example, many cities which applied pilot schemes have set timetables for achieving carbon peaking, thus contributing to structural adjustment; seven provinces and cities, including Beijing, Shanghai, Shenzhen, and Guangdong, have begun to explore the application of market mechanisms to promote low-carbon development; Zhenjiang and other cities have also explored to establish corporate carbon emissions reporting systems and carbon emissions management platforms; Guangyuan has established the Institution for Low-Carbon Development, an institution in charge of actions to promote low-carbon development.

72. Which Cities Have Joined China's Alliance of Peaking Pioneer Cities? Which Year Has Each Member City Set for Achieving Carbon Peaking?

Achieving carbon peaking by 2030 requires cities to take actions first. In 2010, cities contributed to about 60 percent of the total emissions in China. It is predicted that this figure will increase to about 80 percent by 2030, and cities will be the main driving force to achieve carbon peaking and the green and low-carbon transition in China.

In 2015, China and the United States jointly held the first U.S.-China Climate-Smart/Low-Carbon Cities Summit, which was attended by many provinces and cities from China. They announced their determination to make efforts to achieve carbon peaking and the establishment of China's Alliance of Peaking Pioneer Cities (APPC), with the aims of strengthening summaries and sharing of experience in achieving carbon peaking through low-carbon development and emissions reduction, promoting excellent practices of carbon emissions reduction at home and abroad, and playing an exemplary and leading role. 11 provinces and cities joined the alliance as the first batch of APPC members. In 2016, at the second U.S.-China Climate-Smart/Low-Carbon Cities Summit, another 12 cities of China joined the alliance, promising to achieve carbon peaking ahead of 2030, the year set for achieving carbon peaking nationwide. The 23 member cities in the alliance have about 17 percent of China's total population, 28 percent of China's total GDP, and 16 percent of China's total carbon dioxide emissions. See Table 6.

73. Which Industries or Enterprises Have Proposed Targets for Achieving Carbon Peaking and Carbon Neutrality? What Lessons Can They Learn from Each Other?

China's commitment to achieve carbon peaking before 2030 and carbon neutrality before 2060 has been profoundly affecting the general trend in our economy and industries, and changing people's lives. More than 70 percent of global greenhouse gas emissions come from energy consumption, 38 percent from energy supply sectors, and 35 percent from energy consumption sectors, such as construction, transportation, and industry. Therefore, governments must formulate tailor-made industrial policies in accordance with characteristics of these key areas and industries, in a bid to achieve carbon peaking and carbon neutrality as scheduled. In 2020, the Central Economic Work Conference deployed the task named "Being Well-Prepared for Achieving Carbon peaking and Carbon Neutrality" as one of the eight tasks in 2021. Guiding Opinions on Overall Planning and Strengthening the Work Related to Climate Change and Environmental Protection, released by the Ministry of Ecological Environment in January 2021, stated that tailor-made plans must be formulated in accordance with characteristics of energy, industry, transportation,

Table 6. The years that APPC member cities set for achieving carbon peaking

Year to join APPC	Province/city	Targets for achieving carbon peaking
2015	Beijing	around 2020
2015	Sichuan	before 2030
2015	Hainan	by 2030
2015	Shenzhen	by 2022
2015	Guangzhou	before the end of 2020
2015	Wuhan	around 2022
2015	Guiyang	before 2025
2015	Zhenjiang	around 2020
2015	Jilin	before 2025
2015	Yan'an	before 2029
2015	Jinchang	before 2025
2016	Ningbo	before 2020
2016	Wenzhou	before 2020
2016	Suzhou	around 2020
2016	Nanping	around 2020
2016	Qingdao	around 2020
2016	Jincheng	around 2023
2016	Ganzhou	around 2023
2016	Chizhou	around 2030
2016	Guilin	around 2030
2016	Guangyuan	around 2030
2016	Zunyi	around 2030
2016	Urumqi	around 2030

Note: The first batch of APPC members includes 21 cities and 2 provinces and autonomous regions. The second batch of APPC members are all cities.

Source: The author made this table based on available materials.

construction, and other key fields; clear targets for the steel industry, the building materials industry, nonferrous metals industries, the chemical industry, the petrochemical industry, the electric power industry, the coal industry, and other key industries, and action plans for achieving carbon neutrality must be made. In response, many key industries and enterprises have announced their plans and road maps for achieving carbon peaking and carbon neutrality and begun to take concrete actions on carbon emissions reduction (Tables 7 and 8).

By taking the lead in pledging to achieve carbon peaking and carbon neutrality, key industries and large-scale leading enterprises have set a good

Table 7. Carbon peaking targets for key industries

Industries	Targets
The steel industry	It should achieve carbon peaking by 2025.
The building materials industry	It should fully achieve carbon peaking by 2025; the cement industry and other industries should take the lead in achieving carbon peaking before 2023.
The automotive industry	It should achieve carbon peaking by 2028, ahead of the year (2030) set for achieving carbon peaking nationwide.

Source: The author made this table based on available materials.

example and provided good experience so that other industries and enterprises can learn from and replicate. When formulating plans for achieving carbon peaking and carbon neutrality, key industries should prepare thorough high-level designs, formulate road maps for carbon peaking in accordance with major events of achieving carbon peaking and carbon neutrality, and comprehensively adopt relevant policy tools and measures to continuously promote structural adjustment and to promote green and low-carbon development. When achieving carbon peaking and carbon neutrality, industries and enterprises should actively develop and utilize clean renewable energy technologies, give full play to industrial characteristics and advantages, increase the substitution rate of fossil fuels and natural mineral raw materials, and promote the application of advanced and applicable low-carbon emitting technologies. including technologies for ultra-low emissions and carbon capture, utilization and storage. Key industries should integrate efforts to achieve carbon peaking with supply-side structural reforms and promote green, low-carbon, and sustainable development by reducing and eliminating backward production capacity.

74. What Impacts Will China's New Infrastructure Have on Carbon Emissions?

Infrastructure refers to physical engineering facilities that provide public services for social production and citizens' daily life. It can ensure the normal operation of socioeconomic activities of a country or a region and is one of the general physical conditions for the survival and development of society. In 2018, the Central Economic Work Conference first proposed to hasten 5G commercial deployment and to strengthen construction of new infrastructure, such as artificial intelligence, industrial Internet, and the Internet of Things. Subsequently, the concept

Table 8. Targets for key enterprises to achieve carbon peaking

Enterprises	Targets
State Power Investment Corporation	It plans to achieve carbon peaking by 2030.
China Huadian Corporation	It plans to achieve carbon peaking by 2025.
China Datang Corporation	It plans to achieve carbon peaking by 2025, 5 years ahead of schedule, with the installed capacity of non-fossil fuel energy up by 50%.
China Energy Investment Corporation	It plans to hasten the formulation of action plans for emissions reduction to achieve carbon peaking by 2025.
Tongwei Group	It plans to achieve carbon neutrality before 2023.
China Baowu Steel Group	It plans to achieve carbon peaking by 2023 and carbon neutrality by 2050.
Volkswagen Group	It plans to achieve carbon neutrality by 2050.
Xinxiang Bailu Investment Group Co., Ltd.	It plans to achieve carbon peaking by 2028 and carbon neutrality by 2055.

Source: The author made this table based on available materials.

of new infrastructure has been mentioned frequently in relevant documents and conference deployments at the national level. For example, *Report on the Work of the Government (2019)* and a meeting of the Standing Committee of the Political Bureau of the Central Committee of the Communist Party of China Central Committee held in March 2020 emphasized that developing new infrastructure with the new generation of information infrastructure at the core should become one of the important tasks for future economic development. In April 2020, the National Development and Reform Commission clearly defined new infrastructure as an infrastructure system guided by new development concepts, driven by technological innovation, based on information networks, facing the needs of high-quality development, and providing services such as digital transformation, intelligent upgrading, integration and innovation, etc. Existing new infrastructure mainly includes seven key fields: 5G networks, ultra-high voltage power transmission, inter-city high-speed rail and inner-city rail systems, charging stations for electric vehicles, big data centers, artificial intelligence, and industrial Internet. Compared with traditional infrastructure, which includes railways, highways, airports, water conservancy, and other types of major infrastructure, new infrastructure includes information infrastructure, convergence infrastructure, and innovation infrastructure. Information infrastructure includes communication network infrastructure represented by 5G, the Internet of Things, industrial Internet, and satellite Internet; new technology infrastructure represented by artificial intelligence, cloud computing, and blockchains; computing infrastructure represented by data centers, and intelligent computing centers. Convergence infrastructure includes intelligent transportation infrastructure, smart energy infrastructure, etc. Innovation infrastructure includes major science and technology infrastructure, science and education infrastructure, and industrial technology innovation infrastructure. It can be seen that the essence of new infrastructure is information digitalization. New infrastructure can effectively promote economic and social innovation, industrial upgrading, consumption upgrade, and high-quality development. So, it is green, low-carbon, and environmentally-friendly.

In 2020, coal accounted for 56.8 percent of the total primary energy consumption, and clean energy, such as natural gas, hydropower, nuclear power, and wind power accounted for 24.3 percent. That is to say, fossil fuels accounted for more than half of China's energy consumption, and clean energy only accounted for less than a quarter. Construction of new infrastructure, such as ultra-high voltage power transmission networks and charging stations for electric vehicles, can help enlarge the share of clean energy in the total primary energy consumption, promote the utilization of clean energy, and contribute to achieving regional targets for emissions reduction. Construction of inter-city high-speed rail and inner-city rail systems can help enhance the electrification of high-capacity transit systems, improve energy utilization efficiency of transportation systems, alleviate traffic congestion, and significantly reduce air pollutants and carbon dioxide emissions.

Other key fields of new infrastructure are also closely correlated with information and communication technologies. They are also key energy-intensive fields. With the rapid development of the information and communication industry, the scale of communication networks has been continuously enlarged. Construction and renovations of a large number of facilities and equipment have fueled fast-growing demand for energy consumption. In the context of developing new infrastructure nationwide, such as 5G networks, big data centers, and industrial Internet, China needs to utilize a large number of advanced green equipment and technologies, to improve utilization efficiency of resources and energy in the information and communication industry, to accelerate the construction of green infrastructure, and to develop advanced green networks by eliminating old communication equipment with high energy consumption. Only in this way can we effectively enhance the overall efficiency of energy conservation and emissions reduction, achieve green development, and reduce carbon emissions from domestic infrastructure.

75. Is There Any Existing Technology Available for Achieving Carbon Neutrality? Which New Technological Breakthroughs Are Needed?

Scientific and technological innovations are the key to and an important foundation for achieving carbon peaking and carbon neutrality. Therefore, it is necessary to formulate scientific and technological innovation and implementation plans toward that end. Specifically, overall consideration should be given to short-term economic recovery, medium-term structural adjustment, and long-term low-carbon transition; low-carbon/decarbonization technologies should be applied to enhance the competitiveness of green industries in the future; carbon constraints should be listed in The 14th Five-Year Plan for Scientific and Technological Innovation and Development in the context of achieving carbon neutrality before 2060; The Medium and Long-term Special Plan for Science and Technology in Response to Climate Change should be launched in key fields, and relevant studies should be made so that carbon neutrality can be achieved with technical support.

Secondly, it is necessary to speed up the construction of a power generation system in which non-fossil fuel energy will take up a high proportion and to encourage industries to increase the electrification rate. Establishing a power generation and utilization system in which non-fossil fuel energy will take up a high proportion is an important way to ensure carbon neutrality. Specifically, we should encourage technologies of generating electricity from renewable energy and ensure the affordability of renewable energy power generation as soon as possible before 2030. In terms of nuclear energy, we should accelerate research and development and application of new technologies featuring modularity, small size, and differentiation. We should strengthen research and development of energy storage and smart grid technologies and launch more pilot programs so that these technologies can be applied on a large scale by 2040 at the latest, and that non-fossil fuel energy can take up more than 90 percent of the total power generation by 2060. After achieving these two goals, we should encourage industries to increase the electrification rate to more than 50 percent by 2060, which means having cities and towns fully electrified, powering rural areas dominantly with electricity and biomass, having railways basically fully electrified, and increasing the share of electric vehicles in passenger cars to more than 90 percent.

Thirdly, it is necessary to use hydrogen fuel and biofuels as fuels or raw materials towards the energy revolution, and foster research and development of negative emissions technologies. In the fields that are hard to electrify, we should break free from fixed mindsets and adopt revolutionary technologies. For example, the industrial sector should adopt such revolutionary technologies as hydrogen-based steelmaking and bio-based plastics to increase the utilization rate of hydrogen fuel to about 15 percent by 2060; the transportation sector should develop technologies that air transport and marine transport can apply to use biofuels and hydrogen fuel as raw materials, and ensure their large-scale application no later than 2050. To offset greenhouse gas emissions that are difficult to reduce in industrial processes, we should foster research and development of negative emissions technologies. To that end, we should actively promote the development of carbon capture, utilization and storage (CCUS), create road maps for close coupling between CCUS and energy/industry, ensure that CCUS can be widely applied in coal power plants and industry around 2035, and that CCUS can be applied in biomass power generation on a large scale no later than 2045, and hasten research and development of and feasibility studies of direct air capture (DAC), solar radiation management, marine decarbonization projects, and other geoengineering technologies.

Fourthly, it is necessary to strengthen construction of guarantee systems to promote technological research and development and innovation. Specifically, we should formulate road maps and investment plans for research and development of key low-carbon technologies and revolutionary technologies, mobilize industries and markets to foster research and development of low-carbon/decarbonization technologies, and launch pilot programs, so as to create a brand-new innovation-driven system. We should establish state-owned laboratories to develop forward-looking and disruptive technologies and to advance revolutionary core technologies so as to expand new economic growth points in the future. We should establish carbon neutrality demonstration zones in national innovation demonstration zones which have been established to implement the 2030 Agenda for Sustainable Development, so as to promote massive integration and demonstration of low-carbon/decarbonization technologies and low-carbon transition of provinces and cities. We should also actively expand international cooperation and attach importance to the Belt and Road Initiative, platforms for South-South cooperation, and China-EU climate cooperation, so as to deepen transfers and exchanges of low-carbon/decarbonization technologies among countries.

76. What Is the Hydrogen Economy? What Is Green Hydrogen? How Can They Contribute to Achieving Carbon Neutrality?

Hydrogen economy refers to an economic system that relies on hydrogen as industrial raw material and energy, with the aims of achieving zero emissions of greenhouse gases, significantly reducing exploitation of resources on the Earth, and doing less harm to the Earth. Green hydrogen refers to hydrogen fuel produced from zero-carbon energy sources, such as renewable energy and nuclear power.

To achieve carbon neutrality, some sectors that are hard to reduce emissions, such as the industrial sector and the transportation sector, need to invent new production processes and technologies to achieve emissions reduction. Hydrogen is a perfect reducing agent as well as a component of most chemical and petrochemical products. It can also be used as energy without emitting CO_2. Therefore, it can play an important role in deep emissions reduction of the industrial sector and other sectors in the future.

Hydrogen-based direct reduction can be used in steelmaking to produce direct reduced iron. In 2020, a steelmaking system with a capacity of 600,000 tonnes of steel went into production in Ningxia, China. At present, more than a dozen sets of such equipment are under installation in the world. Using hydrogen is the easiest way to produce synthetic ammonia (NH_3), a common chemical product. It is produced when hydrogen reacts with nitrogen after hydrogen is produced. The use of hydrogen can shorten the flow of synthetic ammonia production processes and greatly reduce emissions of smog-related gases. Similar to synthetic ammonia, methanol (CH_3OH) is produced in the process of using hydrogen for reaction (the above-mentioned hydrogen production process can be spared). The research team of the Chinese Academy of Sciences is hydrogenating captured CO_2 to methanol in Gansu, China.

Toluene (C_7H_8) production is also relatively easy and can be produced through existing processes. Ethylene (C_2H_4) can be also directly produced in the process of using hydrogen to react with carbon, though it is under study now.

Hydrogen is produced in the process of using zero-carbon electricity to decompose water. Using existing technologies to decompose water into 1 cubic meter of hydrogen needs 5 kWh of electricity. In the future, this figure is expected to be reduced to 2.8 kWh, and the cost of electrolysis equipment will see a significant decline.

When the cost of photovoltaic power generation drops to less than 0.15 yuan/kWh, the cost of decomposing water into hydrogen to produce above-mentioned chemicals will become truly competitive. It is expected that this will first occur in areas with abundant solar resources in 2025. At that time, industry and sectors will see revolutionary changes.

77. How to Reduce Other Greenhouse Gases Besides CO_2?

Methane (CH_4) is the second largest greenhouse gas after carbon dioxide. It accounts for about 20 percent of global greenhouse gas emissions and contributes about 25 percent to global warming. Although the climate and scope involving the goal of achieving carbon neutrality still remain ambiguous, human beings must make greater efforts to reduce emissions of non-CO_2 greenhouse gases, such as methane, to achieve the long-term goal of tackling climate change.

Sources of anthropogenic methane emissions include coal extraction, oil spills, gas leaks, rice cultivation, rumen fermentation, animal manure management, fuel combustion, landfill, sewage treatment, etc. In recent years, the international community has been increasingly concerned about the issue of global methane emissions reduction. According to the report *Methane Tracker 2021: Helping tackle the urgent global challenge of reducing methane leaks* released by the International Energy Agency in January 2021, it was estimated that the oil and gas industry emitted more than 70 million tonnes of methane into the atmosphere globally in 2020. The contribution of one tonne of methane to climate change is equivalent to that of about 30 tonnes of carbon dioxide. If calculated into carbon dioxide, methane emissions from the oil and gas industry are equivalent to EU's total energy-related carbon emissions. In 2018, Canada and Mexico listed controlling methane emissions from the oil and gas industry as one of their commitments to reduce methane emissions, which had been stated in their nationally determined contributions. In October 2020, the European Commission issued the EU Methane Strategy and took actions to pass legislation in 2021, urging oil and gas companies to reduce methane emissions or leaks.

China is also a major methane emitter. In 2014, China's total anthropogenic methane emissions reached 55.29 million tonnes, equivalent to 1.2 billion tonnes of carbon dioxide, of which methane emissions from coal extraction accounted for 38 percent. Therefore, China has been highly concerned about reducing methane emissions from the energy sector, agriculture, waste treatment, and other sources. Methane is a short-lived greenhouse gas. As long as its emissions are stable, it will not affect the climate system in the long run. On the other hand, emissions reduction means negative carbon dioxide emissions of long-lived greenhouse gases. So, peaking emissions of methane and other non-CO_2 greenhouse gases as soon as possible and continuously reducing emissions can spare us more time to achieve carbon neutrality, and thus are of great significance for carbon neutrality.

78. What Is Carbon Emissions Trading? How Did Global Carbon Markets Evolve?

Carbon emissions trading originated from the Kyoto Protocol adopted in 1997. Also known as carbon trading, it refers to the trading of carbon dioxide emissions permits as a commodity. Based on market economy, it is an important policy tool which can contribute to carbon emissions reduction. Enterprises participating in carbon trading may freely decide whether they will use or trade carbon emissions without exceeding emission allowances. Compared with administrative tools, carbon trading is a more economical and flexible choice for enterprises to reduce emissions.

In 2005, the European Union launched the EU Emissions Trading System (EU ETS). Now 21 carbon emissions trading systems are in operation in the world, including the European Union carbon market, the Korean carbon market, the New Zealand carbon market, and the Regional Greenhouse Gas Initiative (RGGI) of the United States. In 2020, Mexico officially launched trial operation of its national carbon market, marking the debut of the first carbon emissions trading system in Latin America. As of the first quarter of 2021, 18 percent of global emissions had been included in local carbon markets around the world, which had not been interconnected yet.

79. How Does China's Carbon Market Contribute to Carbon Peaking and Carbon Neutrality?

After a long journey, China's carbon market has grown from several pilot programs in seven provinces and cities to a national carbon market. In October 2011, the National Development and Reform Commission released the Notice on Launching Pilot Programs of Carbon Emissions Trading and approved the launch of pilot programs in seven provinces and cities, including Beijing, Shanghai, Tianjin, Chongqing, Hubei, Guangdong, and Shenzhen. In

December 2014, the National Development and Reform Commission released the Interim Measures for the Administration of Carbon Emissions Trading and announced that the national carbon emissions trading market (exclusive for the thermal power industry) would be established by the end of 2017. After several years of trial operation, the Ministry of Ecological Environment had released the Pilot Measures for the Administration of Carbon Emissions Trading, 2019–2020 Implementation Plan for Setting and Allocating Total Allowances in the National Carbon Emissions Trading Market (Exclusive for the Electric Power Industry), and List of Major Emitters Subject to Management of Allowances in the National Carbon Emissions Trading Market by the end of 2020, marking the official operation of the world's largest carbon market. In July 2021, China's national carbon market was scheduled to start online trading, and the first batch of 2,225 power generation enterprises were expected to be included in global carbon markets. At the same time, the seven pilot carbon trading markets were still operating orderly. As of December 31, 2021, a total of 179 million tonnes of carbon allowances changed hands during the 114 trading days, with an aggregate turnover of 7.661 billion yuan and a completion rate of 99.5 percent. The closing price on December 31 was 54.22 yuan/ton, up by 13 percent from the opening price on the first trading day of July 16.

As a market-oriented mechanism for emissions reduction, carbon trading has advantages in saving costs, promoting technological innovations, and bolstering up the enthusiasm of enterprises, compared with traditional financial subsidies and other policies. Carbon markets can also directly finance green and low-carbon development. However, many problems have occurred in the seven pilot carbon trading markets, including a large quantity of allowances left to be traded, low carbon prices, unwillingness of enterprises to trade their allowances, scanty derivatives of carbon allowances, low trading volumes, unwillingness of enterprises and other market entities to engage in investments and financing activities based on carbon allowances, and the inefficiency of the seven carbon markets in terms of their roles and advantages.

China has pledged to achieve carbon peaking before 2030 and carbon neutrality before 2060. To that end, we should formulate strategic plans for achieving carbon peaking and carbon neutrality as soon as possible, clarify targets for national emissions control and mechanisms for allocation of allowances, assign emissions reduction tasks to all levels, and allocate allowances to enterprises and other market entities. Only when we achieve satisfactory results in these aspects, can we effectively trade in carbon markets, give full play to carbon pricing, and innovate carbon financing. In the future, China's carbon market will expand from the electric power industry to the petrochemical industry, the building materials industry, the steel industry, and other industries. Governments will also adopt a series of measures to have it more actively engaged by financial institutions, including fostering carbon assets management companies and professional investors, and developing carbon-related financial products, such as carbon futures. In a word, continuously improving the carbon trading market and forming a reasonable carbon pricing mechanism can contribute a lot to achieving carbon peaking and carbon neutrality.

80. How Does Green Finance Contribute to Achieving Carbon Peaking and Carbon Neutrality?

Green finance refers to any economic activities which support actions on environmental improvement, climate change, and conservation and efficient utilization of resources, that is, financial services which support project investments and financing, project operation, and risk management in the fields of environmental protection, energy conservation, clean energy, green transportation, and green buildings.

China is the first economy in the world to establish a sound policy system for green finance, which includes five pillars: to hasten the construction of a standardized system for green finance; to enhance information disclosure and supervision of financial institutions; to improve incentive and restraint mechanisms; to diversify green financial products and market systems; to deepen international cooperation on green finance. By the end of 2020, China's outstanding green loans hit nearly 12 trillion yuan, the highest in the world; China's outstanding green bonds reached 813.2 billion yuan, the second highest in the world. China is the only country in the world which have set up pilot zones for green finance reforms and innovations. By the end of 2020, outstanding green loans in nine pilot zones for green finance reforms and innovations of six provinces (autonomous regions) totaled 236.83 billion yuan, accounting for 15.1 percent of the total outstanding loans; outstanding green bonds reached 135.05 billion yuan. Green finance has been playing a key role in promoting green development of economy.

To contribute to achieving carbon peaking and carbon neutrality, the People's Bank of China has pledged to focus on achieving carbon peaking and carbon neutrality and other strategic deployments, and to give full play to the three functions of finance in supporting green development, that is, resource allocation, risk management, and market pricing. To that end, it has clarified dozens of key tasks, including establishing long-term mechanisms, improving standards for green finance, and urging financial institutions to make carbon accounting; innovating green financial products and services, and providing professional financing services for participants of carbon emissions trading; preventing climate change-related financial risks and encouraging pilot programs for local financial reforms and innovations; deepening international cooperation and actively participating in global climate governance, to name but a few. To make green finance better serve the goal of achieving carbon peaking and carbon neutrality, the People's Bank of China has also strengthened communication and coordination with key low-carbon industries and sectors, and encouraged financial institutions to provide financial support for photovoltaic glass production projects, wind and solar power generation, and other key fields.

Section IV: The Synergy Between the Goal of Achieving Carbon Peaking and Carbon Neutrality and Sustainable Development Goals

81. How Do Actions on Climate Change Contribute to Reducing Poverty and Inequality?

Poverty eradication and equality are two important goals of the 2030 Agenda for Sustainable Development adopted by the United Nations. Though the world is committed to reducing poverty and inequality, millions of people are still living below basis subsistence level. Threats caused by climate change have been posing new challenges to global efforts to eradicate poverty, and impacts of climate change have been exacerbating inequalities. Many studies at home and abroad have proved that geographical distributions of poverty-stricken areas are highly consistent with areas vulnerable to climate change. As the main areas affected by climate change, poverty-stricken areas may face greater difficulties in poverty eradication as impacts of climate change on agricultural production, water resources, biodiversity, and health have been aggravating ecological vulnerability.

Taking appropriate measures to actively tackle climate change is crucial for poverty reduction and inequality elimination. Impacts of climate change on agriculture have been exacerbating poverty. By taking adaptive measures and strengthening infrastructure construction, poverty-stricken areas can substantially alleviate negative impacts of climate change, such as droughts, rainstorms, and floods, on agricultural production. For example, being vulnerable to climate change, they can introduce food crops with high tolerance to living environment to increase agricultural output, to ensure basic subsistence of local residents, and to reduce poverty; they can establish climate change and natural disasters early warning mechanisms to reduce poverty in a targeted way; they can use modern technologies to monitor occurrences of extreme climate disasters, to enhance their capability of climate disaster prevention and reduction, to effectively avoid deaths caused by extreme climate, and to ensure the development of the agriculture industry, clean energy, and the tourism industry.

While actively adapting to impacts of climate change, poverty-stricken areas can also take some innovative climate change mitigation measures to reduce poverty. They can learn from China's photovoltaic projects for poverty alleviation. For example, they can take advantage of their abundant natural solar energy to help poor households establish distributed photovoltaic systems, which can be wholly funded by governments, or government subsidies plus bank loans and donations, or poor households themselves. After grid connections of clean energy, governments return the proceeds from clean energy power generation to poor households, thus reducing poverty in a targeted way. They can also launch bioenergy development and utilization projects to help poor rural households reduce living costs, increase income, and improve living conditions, thus contributing to energy conservation and emissions reduction. With Chinese characteristics, the aforementioned projects have contributed a lot to tackling climate change and reducing poverty. They have played an important role in reducing the number of poor people and the inequality of regional development, greatly improving the environment of areas vulnerable to climate change or rural areas and reducing impacts of climate change on poverty-stricken areas.

82. How Do the United Nations Framework Convention on Climate Change, and the Convention on Biological Diversity, and the Convention for the Protection of the Ozone Layer Produce a Synergy?

The United Nations Framework Convention on Climate Change was adopted by the United Nations General Assembly on May 9, 1992 amid the United Nations Conference on Environment and Development held in Rio de Janeiro, Brazil, in June 1992, and entered into force on March 21, 1994. Its most important objective is to stabilize concentrations of greenhouse gases "at a level that would prevent dangerous anthropogenic (human induced) interference with the climate system." The Convention stated that "Such a level should be achieved within a time-frame sufficient to allow ecosystems to adapt naturally to climate change." As of the end of June 2016, it had 197 contracting Parties as its members.

The Convention on Biological Diversity is an international convention adopted under the auspices of the United Nations. It was adopted at the seventh meeting of the Intergovernmental Negotiating Committee launched by the United Nations Environment Programme (UNEP) in Nairobi on June 1, 1992. Together with the United Nations Framework Convention on Climate Change, it was signed amid the United Nations Conference on Environment and Development held in June 1992. It entered into force on December 29, 1993. Its objective is to protect endangered plants and animals and to maximize the diversity of the Earth's biological resources for the benefit of present and future generations. It has more than 180 contracting Parties as its members.

The Convention for the Protection of the Ozone Layer, also known as the Vienna Convention for the Protection of the Ozone Layer, is another important international convention on environmental protection launched by the United Nations Environment Programme. It was adopted at the Diplomatic Conference on the Protection of the Ozone Layer held in Vienna, Austria, in March 1985, and entered into force in 1988. Its basic objective is "to protect human health and the environment against adverse effects resulting or likely to result from human activities which modify or are likely to modify the ozone layer" through appropriate international cooperation and action measures. At the Conference on the Protection of the Ozone Layer held in Montreal, Canada, in 2017, about 200 countries signed a new agreement for the protection of the ozone layer.

Climate change, biodiversity loss, and the ozone hole are three serious global environmental challenges for humanity. The world has been increasingly concerned about the synergy between international conventions on climate change, biodiversity, and the protection of the ozone hole. Climate stability is conducive to the continuation of biodiversity, and the ozone layer is an important barrier to all life and climate on Earth. The ultimate objective of the three conventions and three dimensions is to achieve the sustainable development of humanity and the Earth's ecology.

Impacts of climate change on biodiversity and the important role of biodiversity and ecosystems in tackling climate change determine the close relationship between the United Nations Framework Convention on Climate Change and the Convention on Biological Diversity. The United Nations Framework Convention on Climate Change included many issues related to biological diversity, such as land use, land-use change and forestry, mechanisms for reducing emissions from deforestation and forest degradation, and international mechanisms for loss and damage; the Convention on Biological Diversity included the issue of impacts of climate change on biodiversity and other issues and elements related to synergy, mechanisms for reducing emissions from deforestation and forest degradation, and geoengineering.

In 1987, contracting Parties signed the landmark Montreal Protocol under the Convention for the Protection of the Ozone Layer. They pledged to drastically reduce the production and use of chlorofluorocarbons (CFCs) for refrigerants and aerosol propellants, as well as other ozone-depleting substances. In the following decades, all signatories to the Montreal Protocol have phased out nearly 99 percent of ozone-depleting substances. Benefiting from the agreement, hydrofluorocarbons (HFCs) are widely used as an alternative to ozone-depleting substances. But this substance is a greenhouse gas with a very high global warming potential (GWP). To address this problem, Parties to the Montreal Protocol signed the Kigali Amendment in 2016. They pledged to jointly reduce hydrofluorocarbons to achieve the goal of limiting global warming set out in the Paris Agreement. On June 17, 2021, China deposited with the Secretary-General of the United Nations the instrument of acceptance of the Kigali Amendment, which entered into force in China on September 15, 2021.

As both the Convention on Biological Diversity and the Convention on the Protection of the Ozone Layer are closely correlated with climate change, the United Nations Framework Convention on Climate Change has growing relevance to these two conventions in terms of coordination mechanisms and cross-cutting issues. To better achieve the goal of producing a synergy between these conventions, contracting Parties should strengthen collaboration and take joint actions to avoid conflicting objectives and waste of resources and funds. For years of explorations and improvements, essential contradictions and conflicts have not existed between the United Nations Framework Convention on Climate Change and the other two conventions. They have common objectives. So, when formulating national plans to tackle climate change, countries around the world should take into account the contents of the three conventions as well as the three objectives of tackling climate change, protecting biodiversity, and protecting the ozone layer. In the context of actively tackling climate change, the world should strengthen and deepen the synergy between the United Nations Framework Convention on Climate Change and the other two conventions in terms of their institutions and objectives.

83. How Do Nature-Based Solutions Contribute to Achieving Carbon Peaking and Carbon Neutrality?

In 2008, the World Bank released the report *Biodiversity, Climate Change, and Adaptation*, in which the concept of nature-based solutions (NBS) was first mentioned. It emphasizes the importance of biodiversity conservation for climate change mitigation and adaptation. In 2009, the International Union for Conservation of Nature submitted a report to the fifteenth session of the Conference of the Parties to the United Nations Framework Convention on Climate Change, proposing that nature-based solutions refer to "actions to protect, sustainably manage, and restore natural and modified ecosystems that address societal challenges effectively and adaptively, simultaneously benefiting people and nature." It recommended that countries should incorporate NBS into their national plans and strategies to address climate change. In 2015, the European Commission incorporated NBS into its research program Horizon 2020, stating that "Nature-based solutions are actions which are inspired by, supported by or copied from nature." and that "The four goals are: enhancing sustainable urbanization, restoring degraded ecosystems, developing climate change adaptation and mitigation and improving risk management and resilience." These two definitions are slightly different. But simply put, NBS refers to relying on natural forces (such as ecosystems and their services) to deal with a range of challenges and risks including climate change, and it is also a relatively cost-effective solution and approach.

NBS essentially has much in common with the concept of ecological civilization and the concept of harmonious coexistence between man and nature. Compared with emissions reduction measures dependent on technologies, NBS emphasizes ecological connectivity, aiming to use and manage natural ecosystems and inherent laws of natural ecosystems to enhance carbon absorption and storage of land-based natural or artificial ecosystems, including farmlands, forests, grasslands, wetlands, deserts, and oceans; or aiming to improve land use to reduce greenhouse gas emissions. For example, reducing or avoiding deforestation can help reduce carbon emissions; balanced fertilization (targeted fertilization) can help reduce direct and indirect nitrogen oxide emissions from farmlands; peatland protection can avoid carbon emissions.

In 2017, the Nature Conservancy (TNC) and other research institutions proposed a set of Nature Climate Solutions (NCS) at the global level, including afforestation, forest vegetation restoration, fire control and management, coastal wetland restoration, peatland restoration, conservation tillage, etc. They also pointed out that applying nature-based solutions between 2016 and 2030 could contribute 37 percent to achieving the 2 degrees Celsius target set out in the Paris Agreement. At present, agriculture, forestry and other land use accounts for about a quarter of global greenhouse gas emissions, totaling 10–12 gigatonnes of carbon dioxide equivalent per year. Reducing carbon emissions from this field through nature-based solutions is crucial for achieving carbon peaking and carbon neutrality. While helping achieve the goal of climate change mitigation, NBS can also reduce economic losses caused by climate change. NBS-related projects and measures can help increase employment, promote people's livelihood, and reduce poverty, thus contributing to achieving other Sustainable Development Goals. By protecting biodiversity, improving ecosystem functions, and enhancing ecological efficiency, NBS can help human beings establish sustainable food systems, thus ensuring food security and providing healthy diets.

At the 2019 Climate Action Summit, countries and international organizations, co-led by China and New Zealand, worked together and released the Nature-Based Solutions for Climate Manifesto, urging the United Nations to list NBS as one of the nine major action fields to address climate change. Contributions of NBS to the objectives of the Paris Agreement were also recognized in the Zero Draft of the Post-2020 Global Biodiversity. Many countries have taken NBS as an integral part of achieving their long-term emissions reduction targets. Under the guidance of ecological civilization, China has also been playing an active role in leading international cooperation in carbon emissions reduction.

84. Why Is Carbon Emissions Reduction Crucial for Air Pollution Control?

Carbon dioxide and air pollutants, some of which are short-lived climate pollutants (SLCP), are caused by the same source—fossil fuel combustion. Therefore, carbon emissions reduction and air pollution control can produce a strong synergy.

In recent years, China's Air Pollution Prevention and Control Action Plan has produced remarkable results in smog control, thus significantly improving air quality. In 2019, the annual average concentration of PM2.5 in cities at and above the prefecture level in the world was 40 μg/m³, down by 23.1 percent from the 2015 level, overachieving the target set out in 13th Five-Year Plan ahead of schedule. In 2021, air quality of cities at and above the prefecture level was also improved. Compared with 2015 levels, the average concentration of PM2.5 decreased by 34.8 percent, and the concentration of PM2.5 in Beijing decreased by 57.7 percent. Recent years have also witnessed constant improvement in systems for air pollution prevention and control management. For example, price-based incentives for using clean energy for heating, such as coal-to-gas conversion and coal-to-electricity conversion, have played an important role in carbon emissions reduction. But we must be clear that when low-cost measures cannot work any longer, it will be more difficult to control air pollution in the future.

On January 11, 2021, the Ministry of Ecological Environment released the Guiding Opinions on Overall Planning and Strengthening the Work Related to Climate Change and Environmental Protection, requiring coordinated control on emissions from greenhouse gases and pollutants as well as immediate actions on pollution prevention and control and carbon peaking. This means that during the 14th Five-Year Plan period, governments should focus on the goal of achieving carbon peaking and carbon neutrality and improve air quality by reducing carbon emissions with strategic planning, policies and regulations, statistical monitoring, and comprehensive management and control.

85. Why Is the Goal of Achieving Carbon Peaking and Carbon Neutrality So Crucial for China to Participate in Future International Technological and Economic Competition?

Achieving carbon peaking and carbon neutrality requires transformational development of technologies in many industries. Developed countries have basically announced their targets for carbon neutrality, behind which we will see fierce technological and economic competition. As a banner, the goal of achieving carbon neutrality has led countries around the world to invest in research and development of new technologies. In the future, the world will enter an era of technological transformation in energy, the industrial sector, transportation, construction, and other fields. The European Union, the United States, Japan, and other countries have maintained a lead in research and development of new technologies. It is urgent for China to set a clear direction for research and development of new technologies, otherwise China will fall behind again in the new stage of economic development. We do not have much time left. The European Union has set stricter targets for 2030 to make its technologies surpass other countries before 2030.

In the new round of economic and technological transformation, China cannot lag behind, otherwise China's socioeconomic development will be in a very disadvantageous position. A large economy which cannot maintain a lead in technological and economic innovation is difficult to take the lead and become powerful for real.

Since the countries which have maintained a lead in technologies have long been committed to research and development of many zero-carbon technologies, they have more advantages over those which haven't. It has not been a long time since China announced targets for carbon neutrality. Therefore, we have not yet attached much attention to research and development of many transformational technologies which aim for achieving carbon neutrality. In this sense, China has lagged behind other countries. We must act immediately to catch up on technological innovation and economic transformation, which are essential for carbon neutrality. Time seems more urgent for us as the European Union, the United States, Japan, and other developed countries have intensified their efforts to reduce emissions and hasten technology research and development after we announced targets for carbon neutrality. In the context of achieving carbon neutrality, China needs to immediately adjust national research and development arrangements, propose strategic planning for technology research and development in various fields, and direct universities and scientific research institutions to adjust their plans in accordance with special national research and development arrangements. China also needs to clarify the direction of technology research and development for enterprises and urge them to maintain competitiveness in innovative zero-carbon technologies. In the context of achieving carbon neutrality, China's large enterprises have taken technology research and development as an integral part of the pathways for carbon neutrality.

Some very important technologies, such as advanced hydrogen production technologies and nuclear power technologies, need arrangement of national key projects. Especially in the case of nuclear power, China needs to hasten research on the fourth generation of nuclear power and nuclear fusion. We can invest more than 100 billion yuan in research and development of this field in the coming years so as to ensure that China can maintain a lead in the field of energy.

86. How to Promote South-South Cooperation and the Construction of One Belt and One Road in the Context of Achieving Carbon Peaking and Carbon Neutrality?

South-South cooperation refers to all kinds of economic and technical cooperation activities among developing countries (Most developing countries are located in the southern part of the Southern Hemisphere and the Northern Hemisphere, so the economic and technical cooperation among developing countries is called South-South cooperation). It is an indispensable part of international multilateral cooperation which is conducive to development.

In September and October 2013, President Xi proposed the Silk Road Economic Belt during his visit to Kazakhstan and the 21st Century Maritime Silk Road during his visit to Indonesia. These two initiatives are also known as the Belt and Road Initiative. Ways to promote the Belt and Road Initiative include developing economic cooperation partnerships with countries along the Belt and Road and jointly building a community of shared interests, destiny, and responsibility featuring mutual political trust, economic integration, and cultural inclusiveness via existing bilateral and multilateral mechanisms between China and relevant countries, as well as effective regional cooperation platforms. As of February 6, 2022, China had signed more than 200 cooperation documents with 148 countries and 32 international organizations to jointly promote the Belt and Road Initiative.

As most of countries along the Belt and Road are developing countries, international cooperation under the Belt and Road Initiative highly overlaps with South-South cooperation. As an innovative exploration of South-South cooperation, the Belt and Road Initiative has become an exemplification for South-South cooperation. South-South cooperation and the Belt and Road Initiative both involve promoting bilateral or multilateral technical cooperation and economic cooperation among countries, and strengthening exchanges and cooperation in infrastructure construction, energy, the environment, the development of small and medium-sized enterprises, human resources development, health education, and other industrial fields. Among these fields, climate cooperation is one of the key fields and an integral part of China's foreign aid.

China has pledged to achieve carbon peaking before 2030 and carbon neutrality before 2060, to vigorously support developing countries to achieve green and low-carbon development of energy, and not to build new coal power projects overseas. Against such a backdrop, the world has been concerned about China's actions and measures to continuously promote South-South cooperation on climate change and to help developing countries along the Belt and Road to achieve low-carbon development. Compared with developed countries, China, as a large developing country, can be a better exemplification for other developing countries which are desperate to explore how to balance economic development and the goal of achieving carbon peaking and carbon neutrality. The achievements that China has made in promoting low-carbon transition and achieving carbon peaking and carbon neutrality, as well as knowledge, technologies, talents, and comprehensive solutions that China has accumulated in the process, will all play a more important role in promoting South-South cooperation and the Green Belt and Road Initiative.

We should coordinate the national strategy of achieving carbon peaking and carbon neutrality with South-South cooperation on climate change and the Green Belt and Road Initiative. For example, we should incorporate South-South cooperation on climate change with developing countries along the Belt and Road into the strategic framework of the Belt and Road Initiative, clarify the fields, methods, and scale of South-South cooperation on climate change, and publicize the concept of green and low-carbon development as well as advanced low-carbon technologies and industries to other countries through South-South cooperation and the Belt and Road Initiative; we should study and formulate lists of key tasks and demands of countries along the Belt and Road to address climate change and to achieve South-South cooperation on emissions reduction, and promote cooperation and joint development in key fields, such as low-carbon infrastructure, low-carbon industrial parks, low-carbon energy, low-carbon transportation, low-carbon buildings, climate finance, low-carbon trade in products and services, low-carbon capacity building, research and development of low-carbon technologies, etc.; we should improve and innovate platforms and mechanisms for climate cooperation, which is an integral part of both the Belt and Road Initiative and South-South cooperation. We should galvanize extensive participation of governments, enterprises, international institutions, social organizations, and other stakeholders via channels such as existing intergovernmental cooperation platforms, the Asian Infrastructure Bank, the Silk Road Fund, and the China South-South Climate Cooperation Fund, as well as some flexible means, such as government assistance and international trade, investment, and financing, etc.; we should make full use of channels and platforms which publicize programs for South-South cooperation on climate change, in a bid to have the international community know more about the progress and achievements that China has made in low-carbon development and China's climate aid.

· 4 ·

THINK BIG AND ACT NOW!

Antonio Guterres, Secretary-General of the United Nations, once said that climate change would persist, even if the COVID-19 pandemic passed one day. Global climate change affects everyone. It requires governments, enterprises, and consumers to act now. In daily life, every one of us can make contributions to reducing emissions from clothing, food, housing, transportation, and other consumer fields, and to achieving the goal of achieving carbon peaking and carbon neutrality.

87. What Is the Earth Overshoot Day?

In 2006, the international civil society organization, the Global Footprint Network first proposed the Earth Overshoot Day. It marks the day when the Earth falls into the ecological deficit, and humanity has consumed all the resources that the planet can produce over the entire year.

The Earth Overshoot Day is calculated by dividing the world biocapacity (the quantity of natural resources generated by Earth that year), by the world ecological footprint (humanity's consumption of Earth's natural resources for that year), and multiplying by 365 days, the number of days in a year. In other words, ecological overshoot occurs when humanity's resource consumption for the year exceeds Earth's capacity to regenerate those resources that year. According to the Global Footprint Network, the Earth Overshoot Day in 2018 and 2019 was July 28 and July 29 respectively (Figure 17). As the year 2020 was severely affected by the COVID-19 pandemic, the human ecological footprint in this year shrank a lot. The Earth Overshoot Day in 2020 landed on August 22, 24 days later than the date in 2019. This marks its first delay over the past few decades. However, with the global economic recovery, the Earth Overshoot Day landed on July 30 and July 28 in 2021 and 2022 again. Due to the availability of data, estimates of the Earth Overshoot Day may not be accurate. But they warn humanity that ecological overshoot may bring about serious consequences, such as resources depletion, ecological degradation, frequent disasters, etc. It is urgent to protect the Earth.

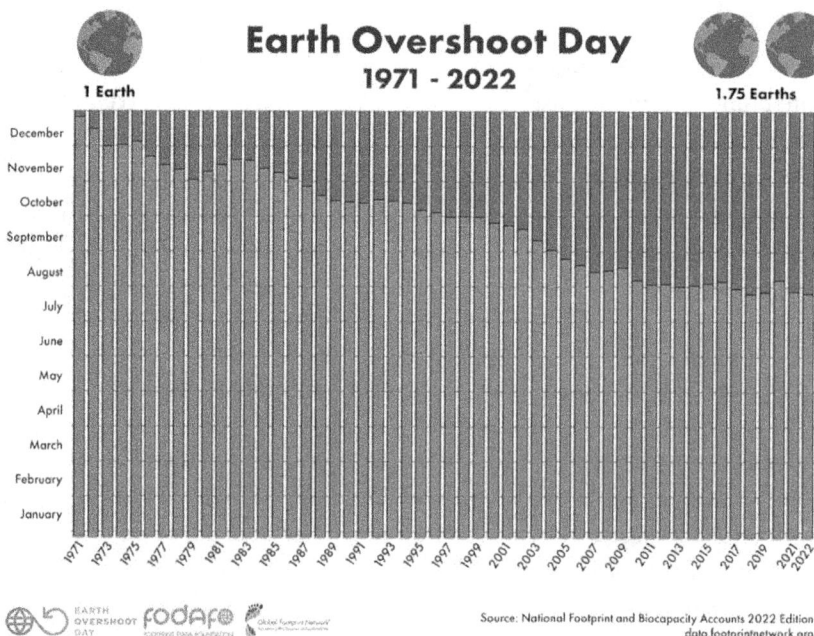

Figure 17. Earth Overshoot Day, 1971–2022
Source: Global Footprint Network National Footprint and Biocapacity Account 2022 Edition, https://www.overshootday.org/.

88. What Can We Do to Achieve Carbon Peaking and Carbon Neutrality?

We cannot live without energy. Many of our daily activities cause carbon emissions. On December 9, 2020, the United Nations Environment Programme (UNEP) released *Emission Gap Report 2020*. In this report, Chapter 6 discusses how to bridge the emissions gap through equitable low-carbon lifestyles. In terms of consumption-based emissions, about two-thirds of global carbon emissions are associated with household emissions; some poor people are still living below basic subsistence level, while some rich people are over-consuming. Emissions from the richest, which accounts for only 1 percent of the global population, are more than twice the total emissions from the poorest, which accounts for 50 percent of the global population. Achieving carbon neutrality requires universal adoption of equitable low-carbon lifestyles. Consumption-based emissions per capita needs to be controlled within 2–2.5 tonnes of carbon dioxide equivalent by 2030 and to be reduced to 0.7 tonnes by 2050. Lifestyle changes are a prerequisite for sustainable emissions reduction of greenhouse gases and for bridging the emissions gap.

Due to the COVID-19 pandemic, millions of people have experienced brand-new low-carbon lifestyles, such as less travel, telecommuting, etc. It can be seen that changing lifestyles rapidly is possible. However, universal adoption of low-carbon lifestyles requires us to make profound changes in socio-economic systems and cultural practices and to create some preconditions for lifestyle changes by improving infrastructure which supports behavioral changes, enhancing services which can make our daily life more convenient, and providing incentives, information, and multiple choices for green and low-carbon lifestyles.

As consumers, we should take actions from the following aspects: 1. We should strengthen the cognition and awareness of carbon peaking and carbon neutrality, and start small in our daily life; 2. We should try our best to obtain information and to know about direct emissions and indirect emissions caused by human daily activities as well as information on energy consumption and emissions of products that we have bought; 3. We should make better consumption choices based on information, including avoiding unnecessary consumption, changing modes of consumption, and reducing carbon emissions from inevitable consumption and its impacts on the environment; 4. We should be well-prepared to pay higher prices for high-quality low-carbon products; 5. We should actively engage in publicity so as to help other people to raise awareness of emissions reduction and to make better choices. Consumer choices can influence producers to make changes, thus greatly contributing to carbon peaking and carbon neutrality.

89. Does Low-Carbon Lifestyle Mean Low Quality of Life?

In the context of achieving carbon peaking and carbon neutrality, we need to adopt low-carbon production, low-carbon lifestyles, and green consumerism. Characterized by nature advocacy and ecological protection, green consumerism refers to a new consumption behavior and process that we moderately control consumption to avoid or reduce the damage to the environment. Advocating green consumerism means at least three aspects: 1. Consumers should choose green products that are not polluted or contribute to public health. 2. While consumers change their thinking on consumption to advocate for nature and to pursue a healthy and comfortable life, they should establish an awareness of environmental protection as well as resources and energy conservation, so as to ensure sustainable consumption. 3. Consumers should properly dispose of waste to avoid environmental pollution in the process of consumption. As the essence of green consumerism, low-carbon consumption emphasizes more on consumption-based emissions reduction. At the beginning of 2020, the Energy Foundation and the *Southern Weekend* jointly released *The Research Report on Low-Carbon Lifestyles and Low-Carbon Consumption Behaviors* through a quantitative investigation of 3,500 respondents from cities above the prefecture level in China and 8 qualitative investigations in Beijing, Hangzhou, Haikou, and Wuhan. It was found that convenient and quick online shopping might contribute to 36 percent of unnecessary consumption behaviors, of which 61 percent were directly associated with online shopping. Low-carbon consumption requires all of us to reduce unnecessary consumption, to avoid impulsive buying, and to resist extravagance and waste.

To live a low-carbon life is to live in a simple, plain, and frugal way. We may need to restrain our impulses, but it does not mean low quality of life. To live a low-carbon life, we need to start small in our daily life, including clothing, food, housing, and transportation. Firstly, we should reduce food waste and save water and electricity. Secondly, we should improve energy efficiency, such as choosing energy-efficient appliances and buildings. Thirdly, we should change our lifestyles, such as taking public transport instead of driving private car. Human needs are diversified. Apart from material needs, spiritual needs are more essential for us, and they cannot be fulfilled by more material consumption. As two traditional virtues of the Chinese nation, living a frugal life and resisting extravagance and waste have influenced the Chinese people to adopt low-carbon behaviors and lifestyles. As our living standards have been constantly improving, we can totally practice the traditional virtue of living a frugal life and the modern value of living a low-carbon life at the same time, so as to live in a healthier, more comfortable, and more convenient way.

90. What Can We Learn from the COVID-19 Pandemic As We Pursue Green and Low-Carbon Development?

Since 2020, the rampant COVID-19 pandemic has triggered a series of unconventional security crises universally and more uncertainties for actions on climate change, an issue that is widely concerned by the world. Studies have shown that nearly two-thirds of new diseases were zoonotic, and over 70 percent of those originated in wildlife. The intensification of climate change has increased the probability of interactions between human beings and animals as well as disease transmissions. The survival and transmissions of coronavirus may be affected by climate conditions, such as temperature, humidity, and aerosols. Low temperature and humid environments are advantageous to the survival of viruses, and high atmospheric stability and high concentrations of particulate matters are advantageous to persistent transmissions of viruses in the atmosphere. However, it remains unknown whether the outbreak of the COVID-19 pandemic is directly associated with climate change. Although measures to control the COVID-19 pandemic have significantly helped reduce emissions of greenhouse gases and aerosols in the short term, their impacts on global climate change are very limited. In 2020, emissions of various greenhouse gases, including carbon dioxide (CO_2), methane (CH_4), and nitrous oxide (N_2O), probably decreased by 5–7 percent compared with the same period, and air quality improved significantly. However, global concentrations of greenhouse gases have not yet seen a significant change. Global warming still persists. According to the report *State of the Global Climate 2021*, released by the World Meteorological Organization, concentrations of major greenhouse gases in the world continued rising in 2021, and the global average temperature was about 1.1 degrees Celsius above pre-industrial levels. According to a report released by the International Energy Agency, in 2021, global CO_2 emissions from fossil fuels totaled 36.3 billion tonnes, and global CO_2 emissions from coal totaled 15.3 billion tonnes; global CO_2 emissions increased by 6 percent, which was in line with the global economic growth rate of 5.9 percent, marking the strongest coupling between CO_2 emissions and GDP growth since 2010. Achieving the 2 degrees Celsius target under the Paris Agreement requires the world to reduce more than 7 percent of global greenhouse gas emissions per year before 2030, a rate much higher than the current level of global emissions reduction.

The COVID-19 pandemic has taught us to keep in harmony with nature. Measures taken by countries around the world to cope with the COVID-19 pandemic and their results have also offered us some valuable clues and experience to study impacts of mitigation measures on long-term emissions, simulation of future emissions scenarios, economic transformation and development, schemes for green economic recovery, etc. If we consider the COVID-19 pandemic a black swan event, global climate change is probably a grey rhino event, warning each country to move towards green economic recovery in the post-pandemic era.

91. What Is Carbon Neutrality? Where Can We Find Carbon Footprint Calculators?

The concept of carbon footprint originates from ecological footprint, which is proposed by researchers from Colombia University. It refers to the total volume of climate change-related gases emitted in human production and consumption activities. Compared with other studies of carbon emissions, carbon footprint analysis is a measure of all direct and indirect carbon emissions associated with all activities in a product's life cycle. Carbon footprint has many different definitions. Scholars from different countries have different understandings of the concept. But generally speaking, it refers to greenhouse gases or carbon dioxide emissions caused by all activities of individuals or other entities, such as enterprises, institutions, activities, buildings, products, etc. It includes direct emissions from fossil fuel combustion in manufacturing, heating, and transportation, as well as indirect carbon emissions caused by goods and services. Carbon footprint can be used to measure impacts of human activities on the environment and to establish a baseline for individuals and other entities to achieve emissions reduction.

Carbon footprint can be divided into national carbon footprint, corporate carbon footprint, product carbon footprint, and personal carbon footprint. National carbon footprint includes greenhouse gas emissions or carbon dioxide emissions from household consumption, public services, and investments. Corporate carbon footprint mainly includes direct and indirect greenhouse gas emissions or carbon dioxide emissions from production activities of enterprises. It is calculated in accordance with ISO 14064, International Standard for GHG Emissions Inventories and Verification, released by the International Organization for Standardization. Product carbon footprint includes greenhouse gas emissions or carbon dioxide emissions from a product's life cycle. It can be calculated in many ways, including PAS 2050: 2008. Specification for the Assessment of the Life Cycle Greenhouse Emissions of Goods and Services (also known as PAS 2050), which is jointly released by the United Kingdom Standards Institute, the Carbon Trust, and the Department for Environment, Food and Rural Affairs in Britain. Widely used in many countries, PAS 2050 is the first standard for product carbon footprint. Personal carbon footprint mainly includes greenhouse gas emissions or carbon dioxide emissions from lifestyles and consumption behaviors of individuals or families.

To calculate personal carbon footprint, many websites have designed their own carbon footprint calculators. Input your daily personal emissions, and you can calculate your carbon footprint (Table 9).

Table 9. Common carbon footprint calculators at home and abroad

Organizations	Websites	Languages
Carbon Footprint	https://www.carbonfootprint.com/calculator.aspx	English
Environmental Protection Agency (EPA)	https://www3.epa.gov/carbon-footprint-calculator/ (Applicable for American households)	English
Conservation International	https://www.conservation.org/carbon-footprint-calculator#/	English
Dotree.com	http://www.dotree.com/CarbonFootprint/	Chinese
Carbonstop	https://www.carbonstop.net/carbon_calculator/standard	Chinese
Beijing Gloriam Climate Technology Co., Ltd	http://www.gloriam.cn/carbonfootprint_cbeex.html	Chinese

92. What Is the Relationship Between Diet Adjustment and Actions on Climate Change?

Human-induced greenhouse gas emissions are the main cause of global climate change. Although multiple studies have shown that industry, transportation, and construction are the main sources of greenhouse gas emissions, food, as the foundation of human survival and development, cannot be ignored anyway. A report released by the European Commission has shown that greenhouse gas emissions from human diets account for 29 percent of the total global greenhouse gas emissions. The agriculture sector is the main source of carbon emissions and the largest source of non-carbon dioxide greenhouse gas emissions. Diet-related carbon emissions include those from the agriculture sector, those embodied in agricultural inputs, and direct and indirect emissions from food processing, transportation, storage, and other links.

With the development of industrialization, urbanization, globalization, and international trade, many countries have witnessed significant changes in dietary structures and lifestyles. Changes in dietary structures are affected by various factors, including income, prices, personal preferences, cultural traditions, geography, the environment, society, economy, etc., and will in turn affect food consumption patterns. Globally, dramatic changes in dietary structures have first emerged in industrialized countries and regions and then spread to developing countries. Previous dietary structures were dominated by fresh and unprocessed plant-based foods, and the total food consumption was relatively low. On the contrary, present dietary structures are dominated by processed animal-based foods rich in sugar and high fat, and the total consumption is higher. According to FAOSTAT, Table 10 lists the changes in consumption per capita of 13 main foods in 1961 and 2011. It is shown that globally, consumption per capita of both animal-based foods and plant-based foods was on the rise, and consumption per capita of plant-based foods was increasing more significantly; dietary energy intake measured by daily calorie intake per capita saw a steady increase.

By comparing the carbon footprint of different foods, many studies have found that the carbon footprint of animal-based foods is much higher than that of plant-based foods. Among protein foods, milk has a lower carbon footprint than eggs. Among meat products, chicken and pork have a relatively low carbon footprint, and beef and mutton have a relatively high carbon footprint. It is noteworthy that mutton has the highest carbon footprint among animal-based foods. One kilogram of mutton emits about 8.34 kilograms of

Table 10. Consumption per capita of 13 main foods in 1961 and 2011
(Unit: kg/p per·year)

	1961	2011
Rice	27.88 ± 39.35	36.78 ± 38.65
Wheat	52.61 ± 51.04	62.32 ± 43.2
Maize	20.83 ± 32.74	23.82 ± 30.06
Legume	6.75 ± 7.58	7.21 ± 7.06
Fruits	62.98 ± 49.99	87.18 ± 54.7
Vegetables	48.01 ± 41.44	80.27 ± 71.05
Beef	11.35 ± 13.16	11.39 ± 9.94
Mutton	4.17 ± 9.39	2.92 ± 4.98
Pork	6.7 ± 9.21	12.7 ± 38.65
Poultry	2.71 ± 3.38	21.49 ± 17.94
Eggs	3.51 ± 4.11	6.34 ± 4.81
Dairy	50.82 ± 60.08	51.34 ± 40.16
Aquatic products	11.82 ± 12.42	20.58 ± 20.77

Source: Yue Wang, "Diet-Related Carbon Emissions in China and Comparison with Those in Foreign Countries" (Ph.D. dissertation, China Medical University, 2019).

carbon dioxide equivalent. Vegetables and fruits generally have a lower carbon footprint. Among vegetables and fruits, cauliflower has a relatively high carbon footprint, and daikon has a relatively low carbon footprint. A kilogram of daikon only emits about 0.014 kilograms of carbon dioxide equivalent, more than 800 times lower than mutton. Among fruits, orange has a relatively low carbon footprint. Among grain crops, maize has a relatively low carbon footprint, and rice has a relatively high carbon footprint. Among vegetable oils, soybean oil has a relatively low carbon footprint, and rapeseed oil has a relatively high carbon footprint. Among sugars, beet sugar has a relatively low carbon footprint, and cane sugar has a relatively high carbon footprint. A study of diets conducted by the University of Oxford has shown that meat-rich diets emit carbon dioxide twice higher than vegetarian diets. One kilogram of meat-rich diets emits 7.2 kilograms of carbon dioxide equivalent per day, and one kilogram of vegetarian diets emits about 3.8 kilograms of carbon dioxide equivalent per day. The energy consumption required to produce animal-based foods is 4–40 times higher than the nutrition when they are consumed. Supposing human beings can take feed grains as food directly, two billion tonnes of food production will be saved. Long-distance movement, processing, and packaging of food commodities will also cause alarmingly huge greenhouse gas emissions and consumption of non-renewable resources.

With the adjustment of global dietary structures, meat consumption per person has exceeded recommended levels globally, and it is expected to continue to rise by 76 percent by 2050. This will pose a serious threat to global health and the environment. The average meat consumption of Americans is three times higher than recommended levels, and that of other industrialized countries is also much higher than recommended levels. This is closely associated with increasing occurrences of obesity, cancers, Type 2 diabetes, and other non-communicable diseases. At present, the average meat consumption of China is lower than that of Western countries. However, meat production, domestic meat consumption, and meat export of China have been continuously increasing. In 1978, China produced 8.5 million tonnes of meat products. In 2011, this figure increased to 79.5 million tonnes, with an annual average growth rate of 6.93 percent. Reducing the intake of animal-based foods in accordance with the principle of keeping a balanced diet and recommended intake levels of various foods can help change dietary structures dominated by animal-based foods and avoid obesity, hyperlipidemia, hypertension, and other chronic diseases caused by excessive food intake, reduce diet-related carbon emissions, and thus contribute to environmental protection. More local foods in dietary structures and reducing the distance of food transportation can also help reduce carbon emissions from transportation. We should also try to choose fresh foods and eat less processed foods, which are detrimental to health and will cause considerable carbon emissions in the process of food production.

93. How Much Carbon Emissions Does Food Loss and Waste Cause?

With the rapid growth of the global population and the continuous decline of the marginal output of agricultural science and technology, food loss and waste has become a global issue. Food production will cause huge consumption of water resources, land resources, agricultural inputs (such as fertilizers, pesticides, etc.), as well as huge greenhouse gas emissions. Diet-related environmental impacts account for 20–30 percent of all types of environmental impacts caused by human waste. So, reducing food loss and waste has become a strategic choice for us to tackle climate change, ensure food security, and protect biodiversity.

According to the Food and Agriculture Organization (FAO) of the United Nations, one-third of the world's food produced for human consumption is lost and wasted throughout the whole supply chain of food production and consumption systems. The report *Food Wastage Footprint: Impacts on Natural Resources* reveals that about 1.3 billion tonnes of food is lost and wasted globally each year, and they emit 3.6 billion tonnes of carbon dioxide equivalent throughout the whole life cycle, similar to the level of the third largest emitter of greenhouse gases in the world. This figure does not include the 800 million tonnes of carbon emissions from forest degradation or organic soil treatment associated with food loss and waste. If all the emissions are added up, they account for 87 percent of global road traffic emissions. Food waste will cause inefficient consumption and greenhouse gas emissions when resources are input in food production as well as huge greenhouse gas emissions when food waste is treated in different ways, thus aggravating climate change. Major greenhouse gases associated with food loss and waste include carbon dioxide, methane, and nitric oxide. The pathways in which food loss and waste aggravates climate change include ineffective chemical inputs in food production and consumption, ineffective emissions associated with aquaculture (such as feed production and processing, rumen fermentation, manure storage and disposal), and food waste disposal (such as incineration, composting, biodiesel, biogas, and feed production). Food waste disposal will produce huge methane and carbon dioxide emissions, and food waste disposal in landfills will cause an 8 percent increase in greenhouse gas emissions. Each year, 100 million tonnes of food are buried in landfills across Europe, and the decomposition and decay of food releases about 227 tonnes of carbon dioxide equivalent, similar to the total fossil fuel emissions of Spain.

China made a late start in studying food loss and waste. Statistics for food loss and waste from various research institutions are quite different. However, it is generally accepted that the problem of food loss and waste in China should be highly concerned. According to Research on Food Waste in China and Other Countries and Advocacy of Patterns of Reusing Food Waste, a program jointly launched by Greenpeace and All-China Environment Federation, the annual food loss and waste in China is 120 million tonnes, which is a rather small statistic compared with other research results. However, this result reveals that food waste on the dining table between 2013 and 2015, if saved, can feed all permanent residents in Beijing and Shanghai. Therefore, reducing food loss and waste will play an important role in achieving carbon peaking and carbon neutrality.

94. What Impacts Will Bottled Water Have on the Environment?

Bottled water refers to drinking water packaged in bottles for personal use and retail sale. For its convenient portability, it has become a popular beverage in modern times. It accounts for more than 10 percent of the total drinking water sales volume worldwide each year. In 2019, China's bottled water sales reached 199.9 billion yuan, and the compound annual growth rate between 2014 and 2019 was as high as 10.08 percent. China has become the world's largest bottled water producer and consumer.

But we cannot ignore the fact that convenient bottled water will cause a large amount of disposable plastic or many packaging products made of other materials. The whole supply chain from water sources to final consumption as well as the final treatment process will all cause great damage to the environment and generate huge carbon emissions.

Manufacturing water bottles will consume a large quantity of resources. Some research institutions have studied the carbon footprint of bottled water, and found that manufacturing bottled water includes many links, such as making plastic bottles, processing water sources, labeling, bottling, sealing, transportation, and refrigeration. Making plastic bottles alone will consume more than 10 million barrels of oil per year. According to a study conducted by the Pacific Institute, a U.S. agency, when all energy inputs are calculated, the energy intensity (energy consumption per unit of output) of bottled water is 5.6–10.2 MJ/L (megajoules per liter of water), equivalent to 2,000 times that of tap water. Materials used to make plastic bottles (polyethylene terephthalate, PET) are generally single-use. It is estimated that making one tonne of PET produces about three tonnes of carbon emissions. Transportation of bottled water after the manufacturing process is also carbon-intensive. Transportation of imported bottled water will consume even more energy than the manufacturing process.

How to dispose of plastic bottles and recycle them after they have been used is another severe environmental problem that we must deal with. Less than half of plastic bottles can be recycled. Only a tiny amount of them can be recycled through official channels as microplastics, which can hardly meet food safety standards. Then most of microplastics will go to the textile industry, the plastic processing industry, and other industries in the form of environment-friendly materials. However, a large number of plastic bottles will be disposed in landfills, waste incineration stations, or simply be discarded directly into ecosystems. Plastic is hardly biodegradable, and non-biodegradation will take a long time. In this process, emissions of harmful gases will affect human health and pollute the environment. Some plastic bottles that cannot be recycled will go into oceans and pollute marine ecosystems. It was found in the process of cleaning up global shoreline pollutants that most marine plastic pollutants were discarded water bottles. If they are mistaken as food by some animals in marine ecosystems, they will pose great risks to their health and survival.

To avoid environmental hazards caused by bottled water, we should try to use reusable water bottles, which can help us protect the environment and save money in the long run. If reusable water bottles are not available, make sure that plastic water bottles will be disposed of properly, and put discarded plastic water bottles into recycling bins after using them, or at least put them into ordinary trash bins if waste sorting is not implemented. Do not discard them directly into ecosystems!

95. How Much Carbon Emissions Does Clothes Waste Cause?

One hundred billion items of clothes are produced worldwide each year. However, 50 percent of them will be discarded within a year. In the United Kingdom, about 235 million items of unwanted clothes were dumped in landfills in 2017, and the amount of old clothes tossed out by Americans is estimated to be 37 kilograms per year. Over the past decades, overconsumption and disposal of unnecessary clothes have become a global concern. Fashion has become one of the most polluting industries in the world. Garment printing and dyeing will consume many chemicals and produce wastewater. Discarded garments will also produce a large amount of garbage and huge greenhouse gas emissions. According to the Alan MacArthur Foundation, global apparel sales volume increased from 50 billion items to 100 billion items between 2000 and 2015, and the number of times an item was worn dropped by 25 percent over the same period. It is predicted that between 2010 and 2030, the global demand for natural and man-made textile fibers will increase by 84 percent; textile production will produce 1.2 billion tonnes of greenhouse gas emissions each year, exceeding the total emissions from all international flights and sea freight. Each year, China produces about 57 billion items of clothes, with about 73 percent of them ending up in landfills.

96. What Is the Significance of Implementing Waste Classification and Management to the Construction of a Low-Carbon Society?

With the improvement of economic development and people's living standard, domestic waste has been increasing constantly. According to the China Association of Urban Environmental Sanitation, domestic waste in cities of China reached 242 million tonnes in 2019. Untreated domestic waste will cause serious pollution to soil environment, atmospheric environment, water environment, and urban environment, as well as energy consumption and carbon emissions in all links, such as garbage collection, transportation, and treatment. For example, as a part of urban domestic waste, kitchen waste will produce huge carbon dioxide emissions in the process of fermentation and decay before going into collection systems, and transportation waste will also produce carbon emissions due to consumption of fossil fuels. Fossil fuels, such as coal, auxiliary fuel oil, and ignition oil, are often added to fuel waste incineration, which will produce carbon dioxide emissions. The combustion of waste itself will also emit greenhouse gases, such as carbon dioxide and nitrogen dioxide, as well as air pollutants. Leachate generated from waste in the storage pit of incineration plants will produce methane and other greenhouse gases in the process of anaerobic fermentation. Conventional landfills will also cause considerable greenhouse gas emissions, such as methane emissions from landfills as well as nitrogen dioxide and other carbon-containing gases emissions from landfill leachate. A series of assessment reports released by the United Nations Intergovernmental Panel on Climate Change (IPCC) clearly stated that carbon dioxide and methane generated in the process of waste disposal have become major sources of anthropogenic greenhouse gases. Therefore, improving the efficiency of waste disposal will effectively help reduce relevant carbon emissions.

Waste classification is a broad concept that encompasses a series of activities that turn waste into public resources through classified storage, collection, and transportation in accordance with certain regulations or standards. By doing so, human beings aim to improve the value of waste as resources and its economic value, and make the best use of it. Common waste can be divided into domestic waste, waste from fairs and other commercial activities, waste in public places, waste from street sweeping, medical waste, and other types of special waste, including solid waste, semi-solid waste, liquid waste, and gaseous waste. Solid waste includes domestic waste and abandoned household appliances; semi solid waste includes sludge and slurry; liquid waste includes waste acid and waste oil, but it does not include wastewater discharged into water bodies; gaseous waste is placed in containers, such as waste gas tanks and waste hydrogen tanks. Construction waste is also a common type of waste. It will cause great damage to the environment, but it is highly recyclable. Sorting, removing or crushing construction waste after classified collection are the best ways to dispose of it so that it can be reused as renewable resources. After the implementation of waste classification, we should dispose of different types of waste properly, so as to increase recycling efficiency and to reduce carbon emissions. Taking the highly recyclable kitchen waste, for example. According to relevant studies, the carbon contained in biogas residues after anaerobic treatment of kitchen waste can be fixed in soils or utilized by plants through land use. Incineration of biogas residues will produce carbon emissions. Burning 1 kilogram of biogas residues will produce 3.6–3.7 kilograms of carbon emissions. If one waste disposal project can treat 800 tonnes of biogas residues per day, burning all biogas residues alone will produce about 100,000 tonnes of carbon emissions each year. On the other hand, waste after being sorted at source can be used to produce green natural gas and to replace fossil fuels, which will contribute to significant carbon emissions reduction.

A low-carbon society refers to a harmonious society in which everyone has the consciousness of low-carbon consumption to live a low-carbon life, to develop a low-carbon economy, and to cultivate a low-carbon cultural concept which advocates sustainable development, environmental protection, and ecological civilization. Waste classification and management can help increase the value of waste as resources and its economic value, effectively improve urban and rural environment, and achieve carbon emissions reduction. So, it is conducive to the construction of China's ecological civilization and low-carbon society.

97. Which Modes of Transport Contribute Most to Green and Low-Carbon Development?

Transportation has become an indispensable part of people's daily life. When people enjoy fast, convenient, and diversified transportation services, carbon emissions from transportation will see an alarming increase. Choosing green and low-carbon modes of transport can effectively help save energy, reduce pollution, and benefit health, but not at the expense of travel efficiency.

Different modes of transport will consume different amounts of energy and produce different volumes of carbon dioxide emissions. According to the Sustainable Transport Research Team of the China Council for International Cooperation on Environment and Development, the energy consumption per capita of buses per 100 km is only 8.4 percent that of gas cars, and the energy consumption per capita of subways per 100 km is about 5 percent that of gas cars. If 1 percent of the population of China shift from traveling by cars to traveling by public transport, 80 million tonnes of fuels will be saved nationwide each year, and 44 million tonnes of carbon dioxide emissions will be reduced, supposing the carbon emissions coefficient of gas is 0.5538.

Based on actual data from a project for Melbourne, Australia's Institute for Sensible Transport measured the average carbon emissions per capita of different modes of transport and studied the space consumption per capita of different modes of transport (Table 11).

It is noteworthy that electric vehicles are not a panacea for tackling climate change. Given the existing capacity of power grids, electric vehicles seem to hold limited emissions reduction potential since they produce only slight carbon emissions than do gas-powered vehicles and take up almost the same amount of space. However, when we consider the entire life cycle of these vehicles, pure electric vehicles may consume as little as 42 percent less energy

Table 11. Carbon emissions and space consumption per capita of different modes of transport in Melbourne, Australia

	Carbon emissions (Grams of CO_2 per person kilometer traveled)	Space requirement (Space in m^2 required per occupant)
Regular sedan	243.8	9.7
The best electric vehicle (Victorian grid)	209.1	9.7
Dual occupancy car	121.9	4.9
Motorcycle	119.6	1.9
Train	28.6	0.5
Tram	20.2	0.6
Gas-powered bus	17.7	0.8
The best electric vehicle (Green energy source)	0	9.7
Bicycle	0	1.5
Walking	0	0.5

and emit as little as 22 percent less carbon than traditional vehicles powered by internal combustion engine. These are significant gains. The electrification of the transportation sector will play an important role in our pursuit of carbon neutrality and the green transition of future electricity systems.

In sum, reducing unnecessary travel is first and foremost in achieving green and low-carbon travel. The development of modern Internet technologies has made video conferencing convenient and fast, thus making it possible to work remotely. The prevention and control of the COVID-19 pandemic have greatly increased the number of people who work remotely and engage in activities online. As a result, the need to travel decreased drastically. For short-distance travel, walking or cycling, which produce no emission, are the best options. For long-distance travel within cities, low-carbon options such as buses or subways are the best. If we have to use private cars, choose electric vehicles or fuel-efficient cars. For inter-city travel, high-speed rail is a good choice.

98. What Role Can Women Play in Achieving Carbon Peaking and Carbon Neutrality?

Gender equality is a measure of social and civilizational progress and one of the important goals in the 2030 Agenda for Sustainable Development. Women play an indispensable role in tackling climate change and achieving carbon peaking and carbon neutrality.

On one hand, women are more vulnerable to adverse effects of climate change than men. In the face of intensifying global climate change, and given their role in society, women will need to do much more in response, and that their livelihood is much more likely to be unstable. In real life, women often lack facilities and technologies that would enable them to cope with climate change and meteorological disasters, as well as opportunities and mechanisms that would enable them to participate in decision-making. Therefore, women are more seriously impacted by climate change and meteorological disasters. As the people primarily responsible for maintaining the livelihood and safety of their family, women make decisions about all aspects of everyday life, including food, clothing, housing, transportation, and others, and they have more knowledge about and better understanding of how to adapt to different climate and environmental conditions. For example, they know what has worked well in their locale in water resources management, soil fertility, restoration of grazing systems, sustainable forest management, and ecosystem-based land management. Indigenous knowledge and folk wisdom can be especially valuable in places with unfavorable environmental conditions and limited availability of official data, and inform tailor-made and feasible adaptation and mitigation measures. Even though women bear the brunt of the impacts of climate change, they also contribute greatly to and benefit greatly from adaptations to climate change. Therefore, they should be empowered with more rights to equal participation and decision making.

With the development of women's cause in China, women have been playing an increasingly prominent role in socioeconomic development. There are more than 50 million women left behind in rural areas of China, and they account for about two-thirds of the rural labor force. Having long been damped by low socioeconomic development, rural areas of central and western China have seen a large number of male labor force migrating to more developed areas for better jobs. Therefore, women left behind at home have to assume the productive role. They are more dependent on natural resources to make a living and thus are more likely to be affected by climate change significantly than men. Chinese women account for about one fifth of the world's female population. Therefore, China's efforts to achieve gender equality in climate actions make great contributions to the progress of the world. Achieving carbon peaking and carbon neutrality needs joint efforts of the whole society. Every one of us should make our own contributions to formulating relevant policies and regulations in the fields of gender equality and climate change, encourage and guarantee women's rights to equal participation, contribution, decision making, and management in the field of climate change by addressing the root cause of gender equality, help women raise the awareness of participating in climate actions, strengthen capacity-building, so as to enable women to contribute more toward carbon peaking and carbon neutrality and earn more opportunities for their own self-development.

99. What Role Can Young People Play in Achieving Carbon Peaking and Carbon Neutrality?

The goal of achieving carbon peaking and carbon neutrality reflects China's determination and ambition to actively tackle global climate change. As the main driving force for the development of the future era, young people should proactively take actions to participate in climate governance and make their voices heard so as to help China achieve carbon peaking and carbon neutrality as soon as possible.

Young people should take the initiative to learn more about climate change. They can seek out ways to learn more about carbon emissions and climate change by taking courses related to the environment and sustainable development, following news and policies, and participating in forums, in order to enhance their understanding of China's goal of achieving carbon peaking and carbon neutrality. They can also share what they have learned with family members and friends both in-person and through online social interactions, which can help more people become aware of carbon emissions and help them establish the awareness of energy conservation and emissions reduction.

Young people should make efforts to reduce their carbon footprints. To that end, they should spend within their means, consume moderately, and practice green consumerism in their daily life. For example, they should choose green and environmentally-friendly modes of transport, such as cycling and public transport, following the "Clear your plate" campaign's call to reduce food waste, voluntarily classify residential waste, improve their diet and eat more vegetables, and try to influence others so they would do the same.

Young people should be proactive in taking climate actions. They are observers in international climate negotiations, and they are also an important target group that the Action for Climate Empowerment aims to help. Young people should integrate their knowledge and ideas into actions on climate governance and express their attitudes by joining societies for environmentally conscious university students and youth delegations of the Conference of the Parties to the United Nations Framework Convention on Climate Change, and engaging in projects of environmental NGOs and climate NGOs as well as scientific research practices of academic institutions in the field of climate change, so as to enhance their leadership in addressing climate change and give full play to their important role in climate actions.

100. What is the Role of the Media in the Pursuit of Carbon Peaking and Carbon Neutrality?

Climate communication refers to social communication activities which aim to tackle climate change by disseminating knowledge about climate change among the public and to find solutions to climate change-related problems by capitalizing on changes in public attitudes and behaviors.

The earliest reports on impacts of human activities on climate change can be traced back to a piece of news report published in the *New York Times* in 1932. With the development of climate change research, climate communication has been separated from environmental communication since 1980s and become an independent field. In recent years, the international community has paid increasing attention to climate change. Climate communication involves disseminating scientific facts about climate change and tracking mechanisms and progress of climate change governance. Recent years have seen an increasing number of excellent climate communicators which have been committed to improving topics, activities, and studies related to climate communication, thus making the importance of climate communication gradually known to governments of various countries. As climate change issues are becoming more complex, climate communication needs to be more professional. Interpretating with professional knowledge in climate communication requires international vision and professionalism, as well as intelligibility and wide acceptability.

The past decade has witnessed the development of climate communication in China. As China becomes an important participant, contributor, and leader in global climate governance, the public is increasingly concerned about climate change-related issues. Apart from Chinese traditional state media, new social media platforms are also covering news about climate change, and representatives from enterprises and NGOs have also frequently made their voices heard in international occasions. Since the seventeenth session of the Conference of the Parties (COP 17) to the United Nations Framework Convention on Climate Change held in Durban in 2011, China has held a series of side events at the China Pavilion successively to demonstrate to the international community China's policies and actions to address climate change, and has made remarkable achievements.

In the context of achieving carbon peaking and carbon neutrality, climate communication is crucial for enhancing public awareness and promoting behavioral changes, and plays an important role in supervising enterprises and governments to implement policies and objectives related to the goal of achieving carbon peaking carbon neutrality.

BIBLIOGRAPHY

"2030 Climate Target Plan." *European Commission.* https://climate.ec.europa.eu/eu-action/clim ate-strategies-targets/2030-climate-energy-framework_en.

"Annual CO_2 Emissions." *Our World in Data.* https://ourworldindata.org/grapher/annual-co2-emissions-per-country?tab=chart.

Bai, Lili. "Zhongguo de gaotie daodi youduo huanbao [How Green Is China's High-speed Rail?]." *China Dialogue,* April 5, 2019. https://chinadialogue.org.cn/zh/4/44158/.

"Butong jiaotong gongju chuxing beihou de zhenxiang ni liaojie ma [Do You Know the Truth Behind Different Modes of Transport?]." *WeChat Public Account, National Development and Reform Commission,* June 8, 2018.

Chai, Qimin, and Moyu Li. "Xinxing jichu sheshi jianshe dui zhongdian hangye tanpaifang de yingxiang pinggu [Assessment of Impacts of New Infrastructure Construction on Carbon Emissions from Key Industries]." In *The 2020 Report on Tackling Climate Change.* Bei-jing: Social Science Academic Press, 2020.

Chao, Qingchen, Changyi Liu, and Jiashuang Yuan. "Qihou bianhua yingxiang he shiying renzhi de yanjin ji dui qihou zhengce de yingxiang [The Evolvement of Impact and Adap-tation on Climate Change and Their Implications on Climate Policies]." *Advances in Cli-mate Change Research* 10, no. 3 (2014): 167–174.

Chen, Ying. "Quanqiu qimei Jincheng qianjing yu zhongguo de yingdui celue [Prospect for Global Coal Phase-out and China's Coping Strategies]." *Environmental Protection,* no. 1 (2019): 20–26.

China National Climate Change Expert Committee and United Kingdom Climate Change Commission. *Zhongying hezuo qihou bianhua fengxian pinggu qihou fengxian zhibiao yanjiu [UK-China Cooperation on Climate Change Risk Assessment: Developing Indicators of Climate Risk].* Beijing: China Environmental Press, 2019.

Deng, Xu, Jun Xie, and Fei Teng. "Hewei tanzhonghe [What Is Carbon Neutrality?]." *Advances in Climate Change Research* 17, no. 1 (2021): 107–113.

Ding, Yihui. *Qihou bianhua kexue wenda [Q&As on Climate Change].* Beijing: China Environmental Press, 2018.

Ding, Yihui. *Qihou bianhua [Climate Change].* Beijing: China Meteorological Press, 2010.

Du, Xiangwan. "Zhongguo tanpaifang zongliang youwang tiqian dafeng [China's Total Carbon Emissions Are Expected to Peak Ahead of Schedule]." *Green Living*, no. 1 (2018): 88–89.

Environmental Defense Fund (EDF). "Baideng qihou xingdong jiedu yi xinxingzhengling [Interpretation of Biden New Executive Order on Climate Actions (1)]." *WeChat* public account, January 28, 2021. https://mp.weixin.qq.com/s/5jSttXRmoSDCPaKQimRIcg.

Feng, Hongli. "Kezaisheng peichunen zhengjie ruhe poju [How to Advance Renewable Energy Storage?]." *Energy*, no. 1 (2021).

"Guanyu dui nifabu jicheng dianlu gongcheng jishu renyuan deng zhiye xinxi jinxing gongshi de gonggao [Announcement on Publicizing the Occupational Information on Technical Personnel in the Field of Integrated Circuit Engineering]." *Official Website of the Ministry of Human Resources and Social Security of the People's Republic of China, the Ministry of Human Resources and Social Security of the People's Republic of China*, January 15, 2021. http://www.mohrss.gov.cn/SYrlzyhshbzb/zwgk/gggs/tg/202101/t20210115_407761.html.

"Guanyu tongchou he jiaqiang yingdui qihou bianhua yu shengtai huanjing baohu xiangguan gongzuo de zhidao yijian [Guiding Opinions on Overall Planning and Strengthening the Work Related to Climate Change and Environmental Protection]." *Official Website of the Ministry of Ecological Environment of the People's Republic of China, the Ministry of Ecological Environment of the People's Republic of China*, January 13, 2021. http://www.mee.gov.cn/xxgk2018/xxgk/xxgk03/202101/t20210113_817221.html.

"Guomin jingji he shehui fazhan shiyiwu guihua gangyao [Outline of the 11th Five-Year Plan for Economic and Social Development]." *China News* online. March 16, 2006. https://www.chinanews.com/news/2006/2006-03-16/8/704064.shtml.

"Guomin jingji he shehui fazhan dishierge wunian guihua gangyao [Outline of the 12th Five-Year Plan for Economic and Social Development]." *The Chinese Central Government's Official Web Portal, Xinhua News Agency*, March 16, 2011. http://www.gov.cn/2011lh/content_1825838.htm.

"Guoxinban juxing lvse jinrong youguan qingkuang chuifenghui [The State Council Information Office of China Holds Briefing on Green Finance]." *Official Website of the State Council Information Office of China, the State Council Information Office of China*, February 9, 2021. http://www.scio.gov.cn/m/xwfbh/xwbfbh/wqfbh/44687/44900/index.htm.

Huang, Lei, Yongxiang Zhang, Qingchen Chao, et al. "Xingbie yu qihou bianhua [Gender and Climate Change]." In *The 2020 Report on Tackling Climate Change.* Beijing: Social Science Academic Press, 2020.

Huang, Lei. "Zhege dongtian youduoleng [How Cold Is This Winter?]." *Encyclopedic Knowledge*, no. 5 (2011): 4–6.

IEA. *Methane Tracker 2021: Helping Tackle the Urgent Global Challenge of Reducing Methane Leaks*. January, 2021. https://www.iea.org/reports/methane-tracker-2021.

IEA. *The Oil and Gas Industry in Energy Transitions*. January, 2020. https://www.iea.org/reports/the-oil-and-gas-industry-in-energy-transitions.

IEA. https://www.iea.org/subscribe-to-data-services/co2-emissions-statistics.

Institute for Sustainable Development and International Relations (IDDRI). Translated by ERR Energy Research Micro-News. *Tianranqi he qihou chengnuo liangge buke diaohe de yinsu [Natural Gas and Climate Commitments, Two Irreconcilable Factors?]*. January 2019. https://mp.ofweek.com/smartocean/a345683820106.

IPCC. "2014: Summary for Policymakers." In *Climate Change 2014: Impacts, Adaptation, and Vulnerability. Part A: Global and Sectoral Aspects. Contribution of Working Group II to the Fifth Assessment Report of the Intergovernmental Panel on Climate Change*, edited by C.B. Field, V.R. Barros, D.J. Dokken, K.J. Mach, M.D. Mastrandrea, T.E. Bilir, M. Chatterjee, K.L. Ebi, Y.O. Estrada, R.C. Genova, B. Girma, E.S. Kissel, A.N. Levy, S. MacCracken, P.R. Mastrandrea, and L.L. White, 1–32. Cambridge and New York: Cambridge University Press, 2014a.

IPCC. "2014: Summary for Policymakers." In *Climate Change 2014: Mitigation of Climate Change. Contribution of Working Group III to the Fifth Assessment Report of the Intergovernmental Panel on Climate Change*, edited by O. Edenhofer, R. Pichs-Madruga, Y. Sokona, E. Farahani, S. Kadner, K. Seyboth, A. Adler, I. Baum, S. Brunner, P. Eickemeier, B. Kriemann, J. Savolainen, S. Schlömer, C. von Stechow, T. Zwickel, and J.C. Minx. Cambridge and New York: Cambridge University Press, 2014b.

IPCC. "2019: Summary for Policymakers." In *Climate Change and Land: An IPCC Special Report on Climate Change, Desertification, Land Degradation, Sustainable Land Management, Food Security, and Greenhouse Gas Fluxes in Terrestrial Ecosystems*, edited by P.R. Shukla, J. Skea, E. Calvo Buendia, V. Masson-Delmotte, H.-O. Pörtner, D.C. Roberts, P. Zhai, R. Slade, S. Connors, R. van Diemen, M. Ferrat, E. Haughey, S. Luz, S. Neogi, M. Pathak, J. Petzold, J. Portugal Pereira, P. Vyas, E. Huntley, K. Kissick, M. Belkacemi, and J. Malley. In press. https://www.ipcc.ch/srccl/chapter/summary-for-policymakers/.

IPCC, 2019: Summary for Policymakers. In: IPCC Special Report on the Ocean and Cryosphere in a Changing Climate [H.-O. Pörtner, D.C. Roberts, V. Masson-Delmotte, P. Zhai, M. Tignor, E. Poloczanska, K. Mintenbeck, A. Alegría, M. Nicolai, A. Okem, J. Petzold, B. Rama, N.M. Weyer (eds.)]. Cambridge University Press, Cambridge, UK and New York, NY, USA, pp. 3-35. https://doi.org/10.1017/9781009157964.001.

IPCC. *Climate Change 2013: The Physical Science Basis. Contribution of Working Group I to the Fifth Assessment Report of the Intergovernmental Panel on Climate Change*. Edited by T.F. Stocker, D. Qin, G.-K. Plattner, M. Tignor, S.K. Allen, J. Boschung, A. Nauels, Y. Xia, V. Bex, and P.M. Midgley, 1–30. Cambridge and New York: Cambridge University Press, 2019.

IPCC. "Summary for Policymakers." In *Global Warming of 1.5°C: an IPCC Special Report on the Impacts of Global Warming of 1.5°C above Pre-Industrial Levels and Related Global*

Greenhouse Gas Emission Pathways, in the Context of Strengthening the Global Response to the Threat of Climate Change, Sustainable Development, and Efforts to Eradicate Poverty, edited by V. Masson-Delmotte, P. Zhai, H.-O. Pörtner, D. Roberts, J. Skea, P.R. Shukla, A. Pirani, W. Moufouma Okia, C. Péan, R. Pidcock, S. Connors, J.B.R. Matthews, Y. Chen, X. Zhou, M.I. Gomis, E. Lonnoy, T. Maycock, M. Tignor, and T. Waterfield, 32. Geneva: World Meteorological Organization, 2018.

Jiang, Dabang, and Yeyi Liu. "Wenshi xiaoying huishi diqiu wendu shangsheng duogao [How High Will the Greenhouse Effect Increase the Global Temperature to?—On Balancing Climate Sensitivity]." *Chinese Science Bulletin*, no. 61 (2016): 691–694.

Jiang, Kejun. "Zhongguo de nenyuan zhuanxing hequ hecong [China's Energy Transition: Where to Go?]." *Power Equipment Management* 40, no 1 (2020):34–35.

Jiang, Kejun, and Xiu Yang. "Zhongguo ditan chengshi tuidong chengshi tanxianfeng shixian tanjingling paifang [China's Low-Carbon Cities: Encouraging Low-Carbon Cities to Achieve Net Zero Carbon Emissions]." *China Environment* 69, no. 4 (2020): 24–27.

Jin, Lina, Yiya Lu, Jingyuan Xie, et al. "Jiyu GREET moxing de xinnenyuan qiche quanshengming zhouqi de huanjing yu jingji xiaoyi fenxi [Analysis of Environmental and Economic Benefits of New Energy Vehicles Produced in the Whole Life Circle Based on the GREET Model]." *Resources & Industry* 21, no 5 (2019): 1–8.Lemoine D. "Abrupt Changes: To What Extent Are Tipping Points a Concern in Coping with Global Change?" PAGES 20, no. 1 (2012): 42. https://doi.org/10.22498/pages.20.1.42.

Lenton, M.T., Johan R., Owen G., et al. "Climate Tipping Points—Too Risky to Bet Against." *Nature* 575, no. 7,784 (2019): 592–595. https://www.nature.com/articles/d41 586-019-03595-0.

Levin, K., and D. Rich. *Turning Points: Trends in Countries' Reaching Peak Greenhouse Gas Emissions Over Time*. Working paper, Washington, 2017. https://files.wri.org/d8/s3fs-public/turning-points-trends-countries-reaching-peak-greenhouse-gas-emissions-over-time.pdf.

Li, Huiming. *Quanqiu qihou zhili zhidu bianqian yu tiaozhan, Yingdui qihou bianhua baogao erlingyijiu [Institutional Changes and Challenges of Global Climate Governance* and *The 2019 Report on Tackling Climate Change]*. Beijing: Social Science Academic Press, 2019.

Li, Junran, and Junjuan Zhao. "Guoji touzi xieding yu qihou bianhua xieding de chongtu yu xietiao yi guoji touzi xieding de shiti guize wei shijiao [Conflict and Coordination between International Investment Agreements and Climate Change Agreements—From the Perspective of Substantive Provisions of International Investment Agreements]." *Hebei Law Science*, no. 7 (2019): 130–142.

Liu, Zhenmin. "Quanqiu qihou zhili zhong de zhongguo gongxian [China's Contributions to Global Climate Governance]." *The People's Daily* online. April 1, 2016. http://theory.people.com.cn/n1/2016/0401/c367652-28244976.html,2016-4-1.

Meng, Guobi. "Lun tanxielou wenti de benzhi jiqi jiejue lujing [On the Essence of Carbon Leakage and Its Solution Pathways]." *Jianghan Tribune*, no. 11 (2017): 128–132.

"Net Zero Tracker." *Climate Watch*. https://www.climatewatchdata.org/net-zero-tracker.

"NDC Registry." *UNFCCC*. https://www4.unfccc.int/sites/NDCStaging/Pages/Home.aspx.

Olivier, J.G.J., and J.A.H.W. Peters. *Trends in Global CO$_2$ and Total Greenhouse Gas Emissions; 2020 Report.* Research report, The Hague, 2020. https://www.pbl.nl/en/publications/trends-in-global-co2-and-total-greenhouse-gas-emissions-2020-report.Pan, Jiahua, Qingchen Chao, Mou Wang, et al. "Houbali shidai yingdui qihou bianhua xinfanshi Zeren gongdan jiji xingdong [A New Paradigm for Addressing Climate Change in the Post-Paris Era: Be Responsible and Take Actions]." In *The 2016 Report on Tackling Climate Change*, 1–17. Beijing: Social Sciences Academic Press, 2016.

Pan, Jiahua, Mou Wang, Ying Chen, et al. "Zhongguo canyu guoji qihou tanpan dingwei yu beidingwei gongping de renshi zhongguo de zeren he gongxian [What Role Should China Play in International Climate Negotiations—An Unbiased Understanding of China's Responsibilities and Contributions]." In *The 2014 Report on Tackling Climate Change*, 1–19. Beijing: Social Sciences Academic Press, 2014.

Pan, Jiahua, and Yan Zheng. "Jiyu renji gongping de tanpaifang gainian jiqi lilun hanyi [Responsibility and Individual Equity for Carbon Emissions Rights]." *World Economics and Politics*, no. 10 (2009): 6–16+3.

Pan, Jiahua, Guiyang Zhuang, Yan Zheng, et al. "Ditan jingji de gainian bianshi ji hexin yaosu fenxi [Clarification of the Concept of Low-Carbon Economy and Analysis of its Core Elements]." *International Economic Review*, no. 4 (2010): 88–101+5.

Peng, Shuijun, and Wencheng Zhang. "Guoji maoyi yu qihou bianhua wenti yige wenxian zongshu shijie jingji [Literature Review on International Trade and Climate Change]." *World Economy* 39, no. 2 (2016): 167–192.

"Qianghua yingdui qihou bianhua xingdong zhongguo guojia zizhu gongxian [Strengthening Actions to Tackle Climate Change—China's National Determined Contributions]." *Xinhua Net*, June 30, 2015. http://www.xinhuanet.com//politics/2015-06/30/c_1115774759.htm.

Qihou bianhua kexue gailun [Introduction to Climate Change Science]. Edited by Dahe Qin. Beijing: Science Press, 2018.

Qixiang yubao yucejuan zhongguo qixiang baike quanshu [China Meteorological Encyclopedia: Meteorological Forecasts]. Edited by the editorial board of China Meteorological Encyclopedia. Beijing: China Meteorology Press, 2016.

"Shixian tandafeng shisiwu shiguanjian [The 14th Five-Year Plan Period Is the Key to Achieving Carbon Peaking]." *Economic Daily* online, January 18, 2021. http://www.gov.cn/xinwen/2021-01/18/content_5580590.htm.

"Shisanwu zhongguo shengtai huanjing zhiliang zongti gaishan [China's Environment Has Seen Overall Improvement During the 13th Five-Year Plan Period]." *China Economic Net* online, October 22, 2020. http://paper.ce.cn/jjrb/html/2020-10/22/content_430474.htm.

Tan, Xiujie, and Shaozhou Qi. "Qihou zhengce shifou yingxiangle guoji touzi he guoji maoyi jingdu chengnuoqi tanxielou shizheng yanjiu [Does Climate Policies Affect International Investment and International Trade?—An Empirical Study of Carbon Leakage During Commitment Periods of the Kyoto Protocol]." *World Economy Studies*, no. 8 (2014): 54–59.

"The European Green Deal." *European Commission*, November 12, 2019. https://eur-lex.europa.eu/legal-content/EN/TXT/HTML/?uri=CELEX:52019DC0640&from=EN.

The Ministry of Ecology and Environment of the People's Republic of China. *Zhonghuaren-mingongheguo qihou bianhua dierci liangnian gengxin baogao [The People's Republic of China Second Biennial Update Report on Climate Change].* December, 2018. http://big5.mee.gov.cn/gate/big5/www.mee.gov.cn/ywgz/ydqhbh/wsqtkz/201907/P020190701765971866571.pdf.

The Preparation Committee of *The Third National Assessment Report on Climate Change. Disanci qihou bianhua guojia pinggu baogao [The Third National Assessment Report on Climate Change].* Beijing: Science Press, 2015.

United Nations Environment Programme. *Emissions Gap Report 2020.* Nairobi, 2020. https://www.unenvironment.org/zh-hans/emissions-gap-report-2020.

Wang, Yanhui. "Guanyu qihou chuanbo celue de sikao [Thinking on Climate Communication Strategies]." In *The 2020 Report on Tackling Climate Change.* Beijing: Social Sciences Academic Press, 2020.

Wang, Shaowu, et al. *Quanqiu biannuan de kexue [Science of Global Warming].* Beijing: China Meteorological Press, 2013.

"Xinshidai de zhongguo nenyuan fazhan [Energy in China's New Era]." *The Chinese Central Government's Official Web Portal, the State Council Information Office of China,* December 21, 2020. http://www.gov.cn/zhengce/2020-12/21/content_5571916.

Xu, Jintao. "Daguo chengnuo yu zhongguo nenyuan moshi de biyao zhuanxing [China's Commitments and Inevitable Transformation of China's Energy Structure]." *Official Website of the National School of Development at Peking University, the National School of Development at Peking University,* November 9, 2020. https://www.bimba.pku.edu.cn/wm/xwzx/xwlx/jsgd/508240.htm.

Yin, Xiong. "Hedian de zhanlue dingwei yu zuoyong [Strategic Positioning and Role of Nuclear Power]." *China National Nuclear Power Network,* February 9, 2021. http://www.heneng.net.cn/index.php/home/zc/infotwo/id/61518/sid/33/catId/162.html.

Zhang, Min. "Ouzhou lvse xieyi yu zhongou qihou bianhua hezuo qianjing [The European Green Deal and Prospect for China-EU Climate Change Cooperation]." In *The 2020 Report on Tackling Climate Change.* Beijing: Social Sciences Academic Press, 2020.

Zhang, Xiaoquan, Qian Xie, and Nan Zeng. "Jiyu ziran de qihou bianhua jiejue fangan [Nature-Based Solutions to Climate Change]." In *Advances in Climate Change Research* 16, no. 3 (2020): 336–344.

Zhang, Yongxiang, Qingchen Chao, Jinghua Li, et al. "Qihou bianhua kexue pinggu yu quanqiu zhili boyi de zhongguo qishi [China's Enlightenment from Scientific Assessments on Climate Change and Global Governance Game]." *Chinese Science Bulletin* 63, no. 23 (2018): 2313–2319. doi:10.1360/N972017-01345.

Zheng, Xiaowen, Yanan Fu, Jiaxuan Zhang, et al. "Yingdui qihou bianhua de qingnian canyu lishi xianzhuang yu zhanwang [Youth Participation in Tackling Climate Change: History, the Status Quo, and Prospect]." In *The 2020 Report on Climate Change.* Beijing: Social Sciences Academic Press, 2020.

Zhongguo qihou yu huanjing yanbian erlingyier [Climate and Environmental Evolutions in China]. Edited by Dahe Qin, Yongjian Ding, and Mu Mu. Beijing: China Meteorology Press, 2012.

"Zhonghuarenmingongheguo erlingyijiunian guomin jingji he shehui fazhan tongji gongbao [Statistical Communique of the People's Republic of China on the 2019 National

Economic and Social Development]." *Website of National Bureau of Statistics, National Bureau of Statistics*, February 28, 2020. http://www.stats.gov.cn/sj/zxfb/202302/t20230203_1901004.html.

"Zhongguo guangfu chanye fazhan luxiantu erlingyier nian ban [China PV Industry Development Roadmap (2020 Edition)]." *Website of China Photovoltaic Industry Association, China Photovoltaic Industry Association and China Center for Information Industry Development (CCID)*, February 3, 2021. https://docs.qq.com/pdf/DUHJuTnlzbG90anBv?&u=a47bc1a36 b844982a078e58eff673954.

Zhuang, Guiyang, and Fan Bo. "Cong ziran zhonglai dao ziran zhongqu shengtai wenming jianshe yu jiyu ziran de jiejue fangan [From Nature to Nature—The Construction of Ecological Civilization and Nature-Based Solutions]." *Guangming Daily*, September 12, 2018. https://epaper.gmw.cn/gmrb/html/2018-09/12/nw.D110000gmrb_20180912_1-14.htm.

Zhuang, Guiyang, and Weiduo Zhou. "Zhongguo ditan chengshi shidian tansuo quanqiu qihou zhili xinmoshi [Pilot Exploration of New Models for Global Climate Governance in China's Low-Carbon Cities]." *China Environment Supervision*, no. 8 (2016): 19–21.

www.ingramcontent.com/pod-product-compliance
Lightning Source LLC
Chambersburg PA
CBHW050641190326
41458CB00008B/2369

9 781636 674230